A Textbook of Forest Entomology

THE AUTHOR

Dr. Sathe, Tukaram Vithalrao (M.Sc., Ph.D., Sangit Visharad, IBT (Seri.), F.I.S.E.C., F.S.E.Sc.) Professor in Entomology, Deptt. of Zoology Shivaji University, Kolhapur has teaching experience of 25 years in Entomology at University P.G. course. He is contributory teacher for P.G. Subjects, Agrochemicals and Pest Management and Sericulture since last 12 years. He has published 165 research papers and 24 books viz., "Preet Patangach" (*Drama*) (1981), Crop Protection from Insects (Marathi) (1991), Sericultural Crop Protection (1998), Biological Pest Control (2000), Cotton Pests and Biocontrol Agents (2001), Insect Pests Predators (2001), Sericulture and Pest Management (2001), Mosquitoes and Diseases (2002), Indian Pest Parasitoids (2003), Agrochemicals and Pest Management (2003), Insect Predators and Pest Management (2003), Shoot Feeding and Sericultural Trends (2004), Vermiculture and Organic Farming (2004), Practicals in Basic Entomology (2005), Fundamentals of Bee Keeping (2006), etc. He guided 15 Ph.D. students and completed 4 major research projects (from CSIR, DST and UGC). He visited Canada (1988), Japan (1988), Thailand (2002, 2004), Spain (2005), France (2005), South Korea (2006), Nepal (2007) for academic work. He is member of editorial board of several journals in India and involved in TV programming of butterflies and moths. He published several popular articles in daily newspapers and delivered several talks through All India Radio. He got several prestigeous awards, "Environmentalist of the year-2003", "International Gold Star", *"Bharat Jyoti Award-2006", "Life Time Education Achievement Award-2006", "Jewel of India", "Education Acumen Award" etc.* He also got fellowships from several scientific societies.

A TEXTBOOK OF FOREST ENTOMOLOGY

Dr. T.V. Sathe

Professor, Entomology Division,
Department of Zoology
Shivaji University, Kolhapur-416 004, India

2024

Daya Publishing House®
A Division of
Astral International Pvt. Ltd.
New Delhi - 110 002

First Published, 2009
Reprinted, 2024

ISBN: 978-81-7035-578-6 (Hardbound)

Published by : **Daya Publishing House®**
A Division of
Astral International Pvt. Ltd.
– ISO 9001:2015 Certified Company –
4736/23, Ansari Road, Darya Ganj
New Delhi-110 002
Ph. 011-43549197, 23278134
E-mail: info@astralint.com
Website: www.astralint.com

Digitally Printed at : **Replika Press Pvt. Ltd.**

Dedicated

to my beloved country

INDIA

Preface

A tremendous progress has been made in past few decades in the field of agricultural entomology. However, the progress in forest entomology in India is almost equal to none except the work of few researchers like Stebbings. Inspite of several colleges and research institutes on forestry no textbook is available on forest entomology. Therefore, efforts have been made through this contribution, although the task is very difficult, to fulfill the gap and demand of students and teachers. Insect pests, diseases and fire are the three worst enemies of forest. Amongst the above, insects rank first in causing damage. Control of insect pests in forestry is very difficult task.

The book contain 22 chapters devoted to description of insect pests with respect to distribution, marks of identification, life cycle, nature of damage, host plants and control measures on highly economically important 20 crops of forest *viz.*, Teak, Deodar, Shisham, Asan, Oak, Willow, Walnut, Silverfir, Mahogany, Poplar, Spruce, Pine, Bamboo, Sal, Babul, Subabul, Neem, Satin and others. Since pesticides lead several serious problems to man and environment, more emphasis is given on biological approaches while suggesting the control measures of forest pests. The book is written in simple language and sketches and photographs are relevantly supporting for understanding the various concepts of pests and their management in forests.

I must acknowledge the help rendered by Ms. Asawari (MBBS), my daughter and Mr. B.V. Jadhav for constructive and photographical work respectively. I hope that the book will be helpful for students, teachers, researchers and silviculturists.

T.V. Sathe

Contents

1

INTRODUCTION

The forest is an ecological system cosisting of three dominated vegetative associations. Forest not only provide timber, fuel, wood, fodder and fiber grasses, nonwood forest products and support industrial and commercial activities, but, also maintain the ecological balance and life support system essential for food production, health and human development. It has been estimated that India's annual woodfuel production is about 287 million m^3. As regards to fuelwood consumption, the World Energy Council estimated that our annual consumption of fuelwood in 2001 was 217 million tonnes, of which only about 18 million tonnes consituted sustainable availability from forests (Mohan Kumar, 2006). Out of 329 million hectares of geographical area of our country, actual forest cover exists only 76.52 million hectares, of this, a significant portion is contributed by lands outside. The actual recorded deforested area/or degraded forest area is much greater. Out of 413 districts the forest cover of which was assessed by the forest survey of India, 1996, 30 districts had no descernible forest cover at all. 150 districts has forest cover of more than 33%; 52 districts, between 19 to 33% and 226 districts, less than 19%.

According to Champion and Seth (1968) India's forest is of four major groups, namely, Tropical, Subtropical, Temperate and Alpine. These major groups are further divided into 16 types.

Indian main forest types include tropical dry deciduous, tropical moist decideous, tropical thorn, tropical wet evergreen, subtropical pine, subalpine, Himalayan moist, temperate, tropical semi-evergreen, montane wet temperate, littoral and swamp subtropical broad leaved hill, subtropical dry evergreen, Himalayan dry temperate and tropical dry evergreen with total % of forest area 38.2, 30.3, 6.7, 5.8, 5.0, 4.3, 3.4, 2.5, 2.0, 0.9, 0.4, 0.2, 0.2 and 0.1 million hectares respectively. Each forest type has some specialized insect biodiversity.

Himalayan region, north eastern part of India, Western Ghats, etc. are the main areas of Indian forest. The western Ghats also known as the Sahayadris.

The rate of losing forest cover in India is 1,44,000 hactares per year. It was slowed down to 24533 hectares during the years 1980 and 1995. According to FAO estimates, the annual deforestation rate in India was 0.6% (0.34 million hectares) during the years 1981 to 1990, implying a loss of 3.37 million hectares over the period (Ravindranath & Hall, 1994). The areas deforested considered of dry deciduos forest is 64%, tropical rainforest 15% and moist deciduous forest was 11%.

Forest Entomology deals with forest insects. It is not only related to the study of forest insects but also to some extent the study of forest trees and the effects of insects on trees and their products. Understanding, the aspects of forest insects such as life cycle, their requirement, their reactions with the habitat and other species of insects and trees, etc. is essential component for forest entomology.

Certain forest types are either susceptible or resistant to insect attack. In forest ecosystem, generally, we find three kinds of insects. The first category had direct effect either on the trees or on trees products produced/derived. The second category insects are natural enemies of first category (pest species) in the form of parasitoids and predators. The third catergory insects are those which live upon the plants of the undergrowth, helping disintegration of waste wood, or feed upon the organic matter in the different layer of forest soil. Such insects have important role in forest ecosystem but, very little is known about them.

The forest trees have different phases *viz.*, nursery/seedling, sapling, vigorously growing young stage, seed stage and death stage and vigorous period between sapling and commercial maturity. Insects are associated with every stage of the growth for causing damage to plants. Even if the tree dies or cut down they are attacked by wood deteriorating insects such as bark beetles, flattened borers, etc. These insects affect the wood by staining and allowing wood for rotting by other organisms. Insects can damage various parts of the tree such as seeds, fruits, flowers, stem, leaves and roots by defoliating or sucking the cell sap or forming the galls or boring the stem or seeds or fruits and mining the leaves. Varieties of insects like aphids, scales, caterpillars, bettles, wasps, etc. cause severe damage to trees. At seedling stage of tree insects such as wireworms, white qrubs, bark beetles, leaf miners, budworms, plant lice, scales, cutworms, root eating insects, etc. cause severe damage.

Indian forest insects have been studied by Stebbing in 1898. He published a compilaton entitled "Injurious insects of Indian forests." Wherein he reported life histories of insects of forest plants of India. In 1918 he also released a book, "Indian forest insects". The fauna of British India volumes are also related to forest insects of India, Myanmar and Sri Lanka.

Accoring to Gamble there are about five thousand different species of trees, shurbs, climbers and bamboos occupied 1/3 of Indian flora. The forest tracts are divided into hot dry region and hot moist region. Some genera and species confined to the hot dry localities or the host moist ones. Some species appear to flourish in both of above. The beetles *Sinoxylon crassum* and *S. anale* attack trees in Indian forest varieties. They act as stem borers. These beetles are serious pets of Sissu *Dalbergia sissoo, Acacia modesta, A. catechu, Sal, Terminalia tomentosa, Albizzia, Pterocarpus, Prosopis,* etc. in western Ghats, Assam, Sind, etc. The beetles *Dinoderus pilifrons* and *D. minutus* are important pests of bamboos all over India. A bruchid, *Caryoborus gonagra* attacks seeds of several forest trees in India. The Lepidopteran noctuid larva *Ingura subapicalis* attack leaves of Sal in Western Ghats, Tarai and Oudh forests in monsoon season. Lasiocampid and noctuid larvae attack sal trees in Indian forest. Several species of lepidopteran larvae specially *Leucoma, Lymantria, Dasychira*, etc. attack leaves of Sal trees as additional individuals. The bark beetle *Sphaerotrypes* is also a bad pest of sal tree in Indian forests. A buprestid beetle, *Sphenoptera aterrima* is destructive pest of deodar. The *Scolytus major, S. minor* and *S. deodara* are also associated with devdar trees in forests. The platypodids, the wood borers, are associated with the trees like deodar, spruce and blue pine. The beetles of genus *Xyleborus* attack the branches of silver fir. The weevils of the genus *Cryptorhynchus* attacks Himalayan pines. Teak, *Tectona grandis* is attacked by the beetle, *Trachypholis hispida* Weber. *Laemotmetus* attacked *Terminalia tomentosa* and *T. beleria,* Sissu and mulberry are attacked by *Anthocomus* sp. Babul *Acacia arabica* and *T. grandis* are attacked by *Psiloptera fastuosa* Fabr. Babul is also attacked by Lepidopterous pests like *Euproctis scintillans, Callitera (Dasychira) grotei* and *Inderbella quadrinotata.*

The Jaman and Chirpine are susceptible to buprestid beetle *Capnodis indica. Tribolium ferrugineum* damages Sal *Shorea robusta. Tribolium confusum. T. castaneum* damage bamboo *Dendrocalamus stictus. Aeolesthes holosericea* damages sal, *Terminalia, Acacia arabica* and mango *Mangifera indica.* Cerabycid beetles *Batocera* spp attack Indian rubber *Ficus elastica,* Oak *Quercus grifithii,* Mango etc. Anthribid beetles damage bamboo trees. While *Ceocephalus* to semul. Many species of *Myllocerus* spp *(weevils)* are associated with Teak. Acacia, Dalbergia and Mango, etc.

Pest diseases and forest fire are the worst enemies of forest trees. They cause greatest damage to forest. Defoliators becomes more dangerous to trees at mature stage than that of full vigor of youth likely, bark beetles too.

Forest insects are successfully controlled now. We should forget the old view of foresters and lumberman that "Forest pests couldn't be controlled". The Silviculturist and Entomologist should pay proper attension towards protection of forest trees at every stage. For protection of forests one should know enough about life cycle and other life activities of the insects and their relationship with forest trees and environment.

It is undoubtedly true that insects cause severe damage to forest trees. Although accurate losses of forest trees due to insects have not been accounted in India. Infact, it is difficult to count. The record of waste due to bark beetles is best example reported from USA. The losses in saw timber in USA reported in 1952 by bark beetles due to defoliation were 4530, 5410 and 30 and 1310 million DFT as mortality and growth impact respectively. It is belived that insects kill twice as much timber than disease causing organisms and seven times the amount killed by fire. The mortalities in forest trees due to bark beetles vary with season and or region.

Damage detection in forest trees is must for adopting appropriate control measures. Indian forest department must pay the attention towards this important fact of pest management and forest protection.

In tropical, subtropical and temperate regions termites are quite injurious to unprotected wooden structures. Many Coleopterous and Lepidopterous insects are potential destructors of forest trees. Thus, insects are important economic factor of our forest industry. Therefore forest protection from insects should not be ignored but planned by high skill.

The forest entomology has very wide scope. It includes understanding of biological phenomenon of forest life for managing forest wealth for the interest of human being and nature. The forest entomology, thus deals with forest insects and their relationship with forest ecosystems and their control.

The forest insects are controlled by preventive and curative methods. Both the methods have significant role in protection of forest trees from insects. Preventive control measures are based on the knowledge of insects and forest environment. Without understanding a tree, the entomologist will not apply the control measures for the insect pests. Therefore, correct identification of trees and forest insects is must for best results of forest protection.

Insect taxonomy is essential for correct identification of genera, species and other groups of nomenclature, and classifying insects and understanding their relationship and origin. The above information will add great relevance in adopting appropriate control measures. The closely related species may be controlled by adopting similar strategies of pest management. The studies related to morphology, biology, behaviour, physiology have also tremendous importance in designing control measures of pest species. Specially, against chemical control since chemical control leads several problems such as air, water and soil pollution, health hazards. killing of beneficial insects, pest resistance, secondary pest outbreak, pest resurgence, interruption to ecocycles, etc. Again chemical control is difficult to adopt in forest ecosystem due to the tallness of trees. Therefore, biological and ecological pest control approaches have tremondous importance in pest management. Infact, finding of an alternative for chemical control is the need of the day for keeping environment safely and ecofriendly. However, the problems of forest pest control are more complex than agriculture and need interdisciplinary approaches from specialized persons from pathology, ecology, chemistry, physiology and taxonomy, etc.

Literature on forest entomology in India is very scanty. Developed countries of the world are already actively engaged in generating the literature and information related to forest pest management. The important publications on forest insects published at global scenario refers to Zoological records (Annual publication by Zoological Society of London),' "Index literature Entomologicae" by Hagen *et al.* (1964)," "Journal of Forestry" (European literaure started in 1902), "Forest quarterly" (provides bibliography and list of literature), Review of Applied Entomology (Published by CIE, London), "Experiment Station Record" (Published by USDA, from 1928), "Forestry Abstracts" (published by CFB, Oxford, from 1940), "Annual Review of Entomology", "Biological Abstracts" (published from 1926). Some more important periodicals in forest entomology refers to "Journal of Forestry", "Forestry chronicle", *Z. ang. Entomology* (West Germany), "Indian Forester" (India), Entomophaga (France) and official publications of Govt. of Australia and South Africa. Literature on forest entomology is rich at global scenario but it is in scattered form. Libraries and internet play an important role in circulating the literature and popularizing the concepts of forest entomology.

Historical Account

The forest entomology had its beginning in the nineteenth century. At global scenario America and European countries paid more attention towards forest entomology. But, very little attention was paid by developing countries, specially, India. From Europe, Germany was the first country to develop foestry. German's developed necessary methods of forest tree protection. However, very scanty and rare publications were available on forest insect pests during the years 1773 to 1798 for control strategies.

Linnaeus suggested preventive control measure that, "freshly cut logs be floated in water for preventing injury from insects." No forest entomology was in existance previous to 1800. No specialized entomologists were there to suggest control measures. Non-specialized men (A man whose primary interest was in other subject than entomology) were conducting forest pest management task (strategies). In 1804 and 1805, Beckstein and Scharfenberg published work related to forest insects in two volumes, was available to use for 30 years. The above period is visualized as early period of development of forest entomology in Europe. It is believed that European forest entomology has four distinct periods of development viz., early period, natural history period, taxonomic biological period and modern period.

During the natural history period following events of development took place. During this period Ratzeburg, the father of forest entomology contributed on forest pathology and forest entomology in a book, "Die Forstinsekten". Its first volume was published in 1837 and second and third in 1840 and 1844 respectively. A handbook. "Die Waldver derber and ire Feinde" published in 1869 was also quite famous.

Ratzeburg published several articles and books, "Ichneumoniden der Forstisekten", published in three volumes (1844, 1848, 1852) was the best work of him on parasitic Hymenoptera. He died in 1871. The valuable work contributed by other workers to this period refers to Kollar, Hartig and Nordlinger from Germany and Parris from France. Kollar contributed on "Treatise on Insects injurious to gardens, Forests and Farms" while, Perris contributes on "Histoire des insects du pin maritime." The above contribution is in ten parts (1851-1871) published in Annals of the Entomological society of France. From America. "Insects injurious to forest and shade trees" was published in 1890. Later. "Insects affecting park and woodland trees" was published in 1905 by Felt as coloured illustrations.

During taxonoic and biological period, several workers contributed on biological and taxonomical aspects of forest insects. Noteworthy amongst them refers to Eichhoff, who has cleared many misconceived notions related to forest insects including bark beetles. He published a book, "Europaischen Borkenkäfer" in 1881. Many German entomologists and teachers contributed to forest entomology during the later part of the ninteenth century. Altum, Nitsche and Henschel were outstanding workers in forest entomology of Germany. From America, taxonomical and biological work on forest insects was published by Hopkins. Similary, Swaine and Blackman also contributed on above aspects of forest insects from USA.

German scientists made their efforts since the beginning to the end of nineteenth century. During the modern period, German entomologists were leading in forest entomological work. They contributed to pioneering work on forest insects. However, later, the work has been spread to old world where forests are socio-economically important. The work of Escherich was highly apperciated due to his editorship of the Journal, "Zeistschrift fur angewandte Entomologie". Rhunber, Prell, Wellstein, Friendricks, Sharfenberg, Schadl, Schwertfeger, Voute, Franz., etc. have made significant contribution from Sweden, Tragardh and Butovitch have made outstanding contribution to forest insects and their damage to forest crops. Similarly, from Finland Saalas, from England Fisher, Munro and Verley, from, Japan Inoye and from Switzerland Bovey made significant contribution to forest insects. However, the work of Ratbeburg and Eichhoff was dominant over other workers. From America about 1915, very little attention was paid on forest insects but, recently many workers contributed entirely on practical forest entomology which resulted in development of foresty in USA. In Canada the oldest laboratory is located at Fredericton, New Brunswik. In Ontario, forest insect laboratory is also actively engaged in protecting forest trees and controlling the pest insects. Other important laboratories involved in solving the problems of forest insects are located at Winnipeg, Manitoba, Saskatchewan, Alberta and Victoria of British Columbia of Canada.

2

Pests of Teak *Tectona grandis*

Teak *Tectona grandis* is one of the important hardwood tree of Indian forest which has commercial utilization in making furniture and other domestic usage. Tamil Nadu, Kerala, Karnataka, Andhra Pradesh, Maharashtra, M. P. and U. P. etc are major teak growing states of India. In India it grows naturally in about 9 million hectares. The teak plant is attacked by about 174 species of insects. The important species of which are listed below.

Sr. No.	Common Name	Scientific Name	Family	Order
1.	Teak weevil	*Astycus lateralis*	Curculionidae	Coleoptera
2.	Myllocerus weevil	*Myllocerus viridanus*	Curculionidae	Coleoptera
3.	Curculionid weevil	*M. carinirostris*	Curculionidae	Coleoptera
4.	Curculionid weevil	*M. discolar variegatus*	Curculionidae	Coleoptera
5.	The cerambycid beetle	*Aegosoma costipene*	Cerambycidae	Coleoptera
6.	The kulsi teak borer	*Stromatium longicorne*	Cerambycidae	Coleoptera
7.	The longhorn teak borer	*Gelonaetha hirta*	Cerambycidae	Coleoptera
8.	The colydiid beetle	*Trachypholis hispida*	Colydiidae	Coleoptera
9.	The cucujid beetle	*Silvanus advena*	Cucujidae	Coleoptera
10.	The dermestrd beetle	*Eugonius gratus*	Endomychidae	Coleoptera
11.	The malacodermid	*Plateros dispallens*	Malacodermidae	Coleoptera
12.	The malacodermid	*Plateros* spp.	Malacodermidae	Coleoptera

contd....

Sr. No.	Common Name	Scientific Name	Family	Order
13.	The chalicophrin beetle	*Psiloptera fostusa*	Buprestidae	Coleoptera
14.	The chalicophrin beetle	*P. viridans*	Buprestidae	Coleoptera
15.	The elaterid beetle	*Adelocera modesta*	Elateridae	Coleoptera
16.	The tenebrionid	*Mesomorpha villiger*	Tenebrionidae	Coleoptera
17.	The crysomelid	*Aspidomorpha sanctaecrucis*	Chysomelidae	Coleoptera
18.	The teak defoliator	*Attelabus* sp	Curculionidae	Coleoptera
19.	The scolytid beetle	*Cryphalus tectonae*	Scolytidae	Coleoptera
20.	The scolytid	*Xyleborus hagedorni*	Scolytidae	Coleoptera
21.	The teak pinhole borer	*Xyleborus noxius*	Scolytidae	Coleoptera
22.	The teak scolytid	*Xyleborus velatus*	Scolytidae	Coleoptera
23.	The hyblaeid teak moth	*Hyblaea puera*	*Hyblaeidae*	Lepidoptera
24.	The teak pyralid moth	*Eutectona machaeralis*	Pyralidae	Lepidoptera
25	The teak scale	*Monophlebus* sp	Coccidae	Hemiptera
26.	The hepialid catérpillar	*Sahyadrassus malbaricus*	Hepialidae	Lepidoptera
27.	The south teak caterpillar	*Alceterogystia cadambae*		Lepidoptera
28.	The cerambycid beetle	*Stromatium barbatum*	Cerambycidae	Coleoptera (fig : 3, 4)

1) The myllocerus weevil -

Myllocerus viridanus Fabr

Distribution

This pest is distributed throughout the Western Ghats, Mount Stuart forest of Anamalai hills, Walayan forest, Calicut forest, etc in India.

Marks of Identification

The weevils are elongate, small measuring about 5 mm in females and 4 mm in males. They are with dead white body colouration from above and below. Sometimes, the colouration may be pale yellow green. The weevil shows long and black clubbed antennae. The head is prolonged into a short beak or rostrum. It has black eyes. Antennae inserted near the end of some distance of eyes. Males are smaller than females and with elongate front legs. The gruls of the weevilis are whitish and leg less. The pupa is exarate type, whitish and elongated with loosely appressed appendages to its boby.

Life Cycle

Adult female lay her eggs in soil superficially or on the tree bark. A single female can lay about150-350 eggs. The oviposition period ranged from 20 to 90 days. The eggs are whitish and oval. Egg hatching takes place within 4 to 8 days. Newly emerged tiny grub feed on bark or under ground parts of teak. The grub moults for four times. The grub is soft, whitish, curved and legless. It is with strong mandibles and with yellowish brownish head. The full grown grub pupates in soil. The pupa is white, elongate, narrow with soft rostrum. The pupa is exarate type. The pupa moult into adult weevil. There are 3 to 5 generations in single year. This pest is found throughout the year but has peak from july to october.

Nature of Damage

The adults (weevils) are diurnal creatures and feed on the leaves of teak. Habitually the weevils feed in a typical manner that while feeding on the leaves, the weevil keep untouched main veins and consume other green content of the leaf. Very interestingly, the weevil does not eat the parenchyma of the leaf but leave some small patches of unconsumed tissues between the midrib and the veins. Unconsumed short pieces of the tissues are also found on the side veins of the leaf. The leaf edges of some places eaten out of the leaf are ragged and not clean cut by lepidopterous caterpillars. Defoliation to the tree by this pest is very serious. Entire plants are skeletonised by the weevils, and with the help of two lepidopteran caterpillars.

Host Plants

This pest is equally destructive to the other host plants such as apple, cashewnut and mulberry by defoliation.

Control Measures

Preventive : I) Digging the field at the depth of 7-8 cm will expose the eggs, grubs and pupae of the pest for natural mortality factors such as biotic and abiotic.

II) Hand picking of weevils and destruction of them.

Chemical

I) Spray the crop with 0.01% quinolphos or

II) 0.02% methylparathion or

III) 0.02% malathion or

IV) Foliar spray of any one of the following Diamethoate/ carbaryl/dichlorovos/ endosulphan/ monocrotophos/ phosphamidon at insecticidal concentrations.

2) Other weevils

Myllocerus carinirostris Marshall (Fig. 2.1)

Distribution

Konbilin forest, Tharra waddy, Western Ghats, Maharashtra.

Marks of identification (Fig. 2.1)

Adult is blackish or chestnut brown coloured. Sides and under sides are pale metallic green. Prothorax is dark brown with two narrow green stripes dorsally. Rostrum is as long as broad, longer than head, antennal scape reaching the middle of the prothorax. Legs are red brown with green and gray scaling. The weevil measures about 6 mm in body length and 3 mm in width.

Nature of Damage

Weevils feed on leaves and remove the green tissues and by that affect the photosynthesis and vitality of the teak plant.

3) *Myllocerus discolor variegatus* Boh.

Distribution

Western Ghats, Chennai, South Coimbatore, Mount Stuart, etc.

Mark of Identification

Adult is with a very mobile prothorax and convex elytra. Dorsally it is grayish blue with orange brown patches and mottled black centres. Its proboscis is short and blackish on tip, eggs are stout and antennae are blackish gray.

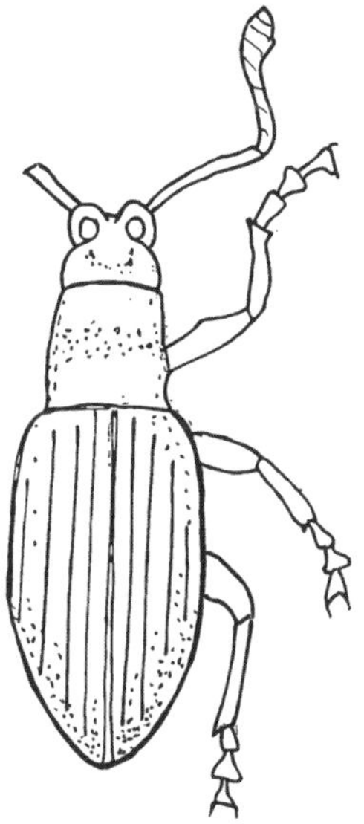

Fig. 2.1 : *Myllocerus carinirostris*

Nature of Damage

Adults are defoliators of teak.

Life Cycle

Similar to above species.

Host Plants

Dalbergia paniculata, Acacia intsia, etc.

Control measures

As above species

4) The teak defoliator weevil

Attelabus sp.

Distribution

Western Ghats, Coimbatore, Chennai, Mount Stuart.

Marks of Identification

Its head, thorax, antennae, legs and under surface are crimson red. Elytra is yellow with 3 dorsal lines. It is moderately shining and measures about 7.5 mm in body length. Its front leg femora is thickened and head is extended into snout.

Life Cycle

Eggs are pale yellowish, elloptical and shining and deposited on *Helicteres* leaves. Egg hatching takes place within 3-6 days. The qrub is whitish and typically tapered towards both end. Pupa is brownish and exarate type.

Nature of Damage

Weevils cause damage to teak leaves by feeding upon them and defoliating in a very typical manner by cutting edges of the teak leaves and preparing slits under attack. In severe infestation the veins as well as parenchyma is eaten up by the weevil. Thus, the infected leaf shows cut edges patches eaten up by the weevil.

Host Plants

Teak *Tectona grandis* and *Helicteres* sp.

Control Measures

As suggested for *myllocerus* weevils.

5) The kulsi teak borer

Stromatium longicorne Newman (Fig. 2.2)

Distribution

Assam, Kulsi, Western ghats, UP. Out side India, it has been recorded from Myanmar and China (South Part).

Marks of Identification

The beetle measures about 17 to 25 mm in body length and 4.5 to 7.5 mm in width. There is much variation in size of the beetle. It has testaceous to dark brown colouration. It is with densely greyish tawny pubescence. Antennae are nearly about the twice the length of body in males of little longer than body in females. First segment of antenna closely punctate at the base in males.

Prothorax is strongly and densely punctate and some what rounded in males and more sharply in females and with two rounded black spots. Elytral apex is rounded. Elytra is armed with sutural spine. The qrubs are whitish yellow in body and with dark brownish head.

Life Cycle (Fig. 2.2)

Eggs are rounded. They hatch within a week. Eggs are laid on the stem of teak. Newly hatched qrub start boring into the stem and feed on woody content. The qrubs are more dangerous to young teak plants. They cause considerable damage to young teak by boring stem. The qrubs are noticed through out the year infesting trees.

The qrub is tuberculated which measures about 35-37 mm in body length in full grown form. It is thick set with strong and black mandibles for boring the stem of teak. The qrub has light brown shining prothorax which is broder than abdomen. It tapers posteriorly. The last two segments of abdomen are elliptical in shape. The larval period is about 3-4 months. The qrub prepare small punctures in the bark for aeration in chamber/bored tunnel. The wood fibre and excreta of qrub many times blocks such aeration. The full grown qrub prepares rough semicocoon with the help of wood fibre particles and dried particles of frass/excreta and pupate in it. The pupa is brownish and exarate type. This stage lasts for two months. Thus, life cycle is completed within 6 months and only two generations are possible in a year.

Nature of Damage

The newly emerged qrubs bore into the stem resulting swelling of the stem just above the ground, many times about 2 feet above the ground level. Several punctures are observed in the bark of swellon portion of the stem. Many times larval excrement is found oozing out from those exits. The infested trees are snapped of by the wind current. The swelling is resistant action of plant against damage which allow the plant to survive successfully in many trees. Boring affects the vigour of the plant and thus, the quality of wood in case of teak plant. Very interestingly, the grub is found at a little above the swelling eating the pith of the tree for several inches in an upward direction. The pith of cavity of swelling is also eaten up completely by the qrub. One to two year old trees are more succeptible to pest attack than five to six year old trees.

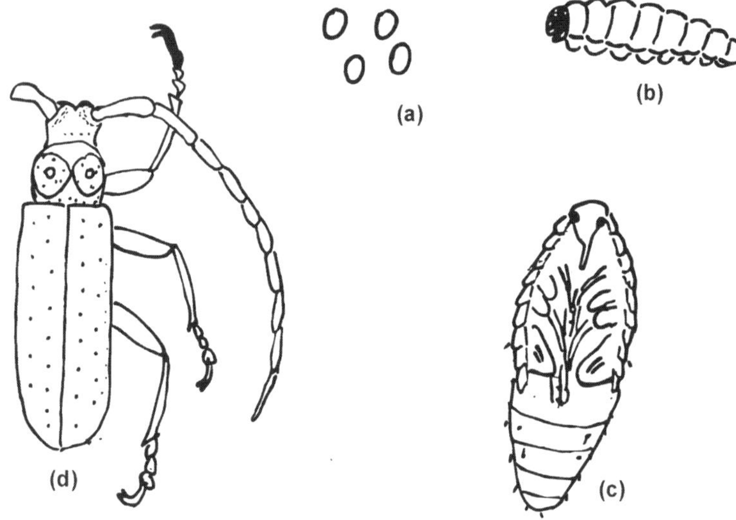

Fig. 2.2 : ***Stromatium longicorne :*** **Life cycle : a-eggs, b-larva, c-pupa, d-adult**

Host Plants

Teak *Tectona grandis*

Control Measures

1) Cutting infested trees and removal of infested parts.
2) Cutting infested trees at younger stage and encouraging growth of a single strongest shoot on that cut down plant.
3) Collection and destruction of beetles when they emerge from teak plant in monsoon season for mating and egg laying.
4) Pouring of kerosinized water/petrol/chloroform soated balls (cotton) in bored tunnel and sealling with mud will kill the bores inside the the tunnel.
5) Insecticide treatment : Any ecofriendly insecticide/ dieldrin/aldrin be used for control of teak borer.

Fig. 2.3 : ***Stromatium barbatum*** (Adult)

Fig. 2.4 : ***S. barbatum*** (qrub) with bored stem.

6) Long horned teak borer

Gelonaetha hirta Fairm.

Distribution

Kolkata region, Nigiri hills and Western Ghats of India. Out side India, it has been recorded from Phillippine islands, Sandwich island and Tahiti etc.

Marks of Identification

The beetle measures about 10 to 16 mm in body length and 2.5 to 4.5 mm in breadth. It has reddish brown to dark brown body which is clothed with greyish pubescence. Head is with feeble median groove. Prothorax is rugulose punctate, elytra is densely punctate and with one or two greyish longitudinal lines. Abdominal tip is extended out of the covering of elytra.

Life Cycle

The beetles are seen in march flying in the forest environment. The female lays oval eggs on the stem of teak plant. The eggs hatch within 8 days. Newly emerged qrub start girdling the stem and killing poles of the teak. The qrub bore and feed in the bast and sapwood preparing narrow winding galleries. Later, they are blocked due to wooddust formed by the pest. The full grown larva bore into the sapwood preparing an elongate narrow chamber which is parellel to long axis of the plant. It pupate inside the gallery for 2 months. The pest can complete two generations in a year.

Nature of Damage

The qrubs are destructive creatures. They girdle the stem resulting killing of poles and later, the full grown qrub bore into the sapwood and tunnel out a narrow chamber as a result, the vitality and vigour of the plant is adversely affected. The larva feed on bast and is potential killer of poles of teak.

Host Plants

Teak *Tectona grandis*

Control Measures

As above pest.

7) The teak pinhole borer

Xyleborus noxius Sampson

Distribution

Tamil Nadu, South Malabar, Western Ghats, Tenasserium, Kowloon Island.

Marks of Identification

The piceous brown beetles are very small, measuring about 2 to 2.5 mm in body length. Their antennae are yellowish brown with yellow club. Prothorax is longer than wide, appex is strongly rounded. Its scutellum is smooth, round and large. Under surface of the beetle is punctate and light coloured with sparse spiny setae. Legs of the beetle are brown coloured, the tibia shows four teeth on outer edge anteriorly and a terminal hook. Elytra is approximately as wide as thorax at base.

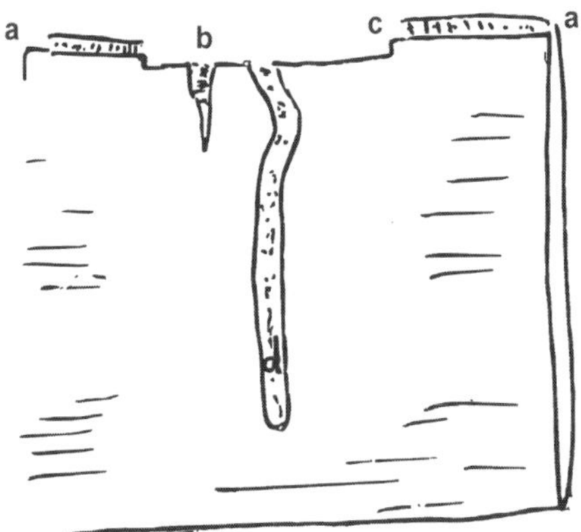

Fig. 2.5 : ***Xyleborus noxius :*** **damage wood of** ***Dabergia cultrata.*** **a-bark, b-c- transverse gallery in outer sap, d-egg gallery.**

Life Cycle

In bark gallery the male fertilizes the female. The mated female lays about four eggs in each tunnel on nearly dead and newly dead trees of teak. The female tunnels down through the bark up to sap wood and lay their eggs at the bottom of tunnel. The female start tunnelling from mating chamber or mating gallery to wood. The tunnel has a special feature that at first it is curved but later, it becomes straight for about 4-5 inches (Fig. 2.5) in each long tunnel 2 to 4 beetles can be observed. Males are polygamous. In single year several generations are completed by this pest.

Nature of Damage (Fig. 2.5)

The beetles are destructive, they bore into the bark and tunnel the wood of teak, tunnelling affect the quality of wood. The market value of infested wood is adversely affected.

Host Plants

Teak *Tectona grandis*

Control Measures

A) Preventive :

1) Collection and destruction of beetles.

B) Natural :

Encourage predaceous beetles against this pest. *Anthocomus* sp. feed on thus pest and suppress the pest population upto certain extent.

C) Chemical :

1) Filling mating galleries and tunnels with kerosinized water and sealing with muds will provide good control of this pest.
2) Treating crop with any one of the following pesticides when the beetles start emerging from the pupae and ovoposting the trees.

a) Dusting with 2% endosuphan.
b) Spraying with 0.2% BHC.
c) Dusting 10% BHC.

8) The teak scale

Monoplebus sp.

Distribution

South India, Western India, Western Ghats, etc.

Marks of Identification

Scale insects are semicircular or oval in shape and with brownish black body colouration. Nymphs and females are wing less and only males are winged. Males do not feed and are not destructive creatures. Only nymphs and females cause damage to plants by sucking cell sap. Males are short lived insects. They fertilize the females and die.

Life Cycle

Its reproduction takes place by vivipariously. Female gives birth to nymphs. Newly emerged nymphs find their places on the tender leaves and stems of teak plant and settle their for sucking the cell sap from teak plant. The nymph moult for 3 times to become the adult.

Nature of Damage

Nymphs and adults cause the damage to teak plant by sucking the cell sap from tender leaves and stem and affect the growth. The leaves becomes curly, they turn yellow and drop down affecting the growth of the plant.

Host Plants

Teak *Tectona grandis*

Control Measures

1) Scrapping the infested plant parts with wooden knife for collection of scales and further killing by dipping into kerosinized water.
2) Spraying the crop with 0.03% Diamethoate or 0.03% malathion or 0.03% phosphamidon or 0.05% DDVP.
3) Synthetic insecticides like cypermethrin/fenvalerate also found effective.

9) The hyblaeid caterpillar

Hyblaea puera Cramer (Fig. 2.6)

Distribution

South India, Western Ghats, Maharashtra, Dehradun. This pest is also reported from Australia, Myanmar, Sri Lanka, Java, S. USA, Africa, etc.

Marks of Identification

The moths are with greyish brown forewings and hind wings are with black and orange-yellow markings. It has wing span of 3-4 cm. Larvae are defoliators. Pupae are brownish and obtect type.

Life Cycle

Mated female lay her eggs singly near the veins on the under surface of tender leaves. A single female can lay about 400 eggs. However, Sudheendra Kumar (1991) recorded 1000 eggs laid by this pest on the host plant. The eggs hatch within 2 days. Newly emerged tiny larvae feed on tender leaves of teak plant. The larva moult for 5 times and prefers only young and tender leaves of teak for feeding. Larval stage lasts for 10-12 days or may extend up to 20 days in cold weather (November). The full grown larva descend down on the ground with the help of silken threads and pupate in soil or in litter. Depending upon climatic conditions, the pupal period lasts for 4-8 days. Thus, the life cycle is completed within 19-36 days. In cold weather the life cycle period is extended. Many generations are possible in a year.

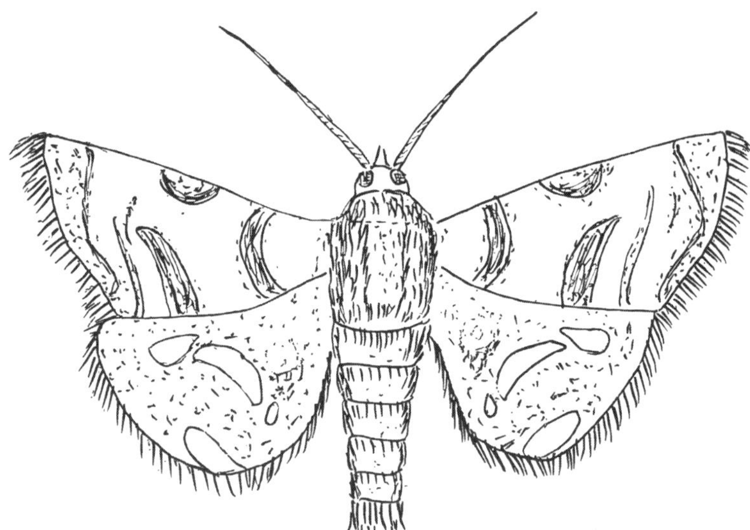

Fig. 2.6 : *Hyblaea puera*

Nature of Damage

Caterpillars are only destructive. They feed on tender leaves of teak and affect the growth of the plant. Damage losses are about 44%.

Host Plants

This is oligophagus pest. Its alternative hosts are scattered in the familes verbenaceae, Bignonaceae, Araliaceae, Jugalandaceae and Oleaceae. However, teak is the principal host of this pest.

Control Measures

1) Collection and destruction of egg mass larvae and pupae.
2) Encourage the following natural enemies for suppression of this pest.

a) Parasitoids

I. *Brachymeria lasus* (Chalcidae : Hymenoptera)
II. *Eriborus gardneri* (Ichneumonidae : Hymenoptera)
III. *Plalexorista* sp. (Tachinidae : Diptera)
IV. *Sympiensis* sp. (Eulophidae : Hymenoptera)

The above parasistoids cause about 70% mortalities in larval stages of the pest.

V. *Bacillus thuringiensis*, NPV also play an important role in suppression of above pest.

3) Clean cultivation, digging the field for exposing pupae for natural mortality factors. (Biotic & Abiotic)
4) Use of Sexpheromones and pheromone traps.

10) The teak pyralid moth

Eutectona machaeralis (Walker) (Fig. 2.7)

Distribution :

Western Ghats, Maharashtra, Dehradun, A.P, T.N., Karnataka, etc. Outside India, it has been recorded from Australia and Indonesia.

Marks of Identification (Fig. 2.7)

The moths are with bright yellow with fulvous or pink transverse marking of zigzag pattern or serrate lines on the fore wing white, hind wings are pale and having an ocherous or reddish marginal line or band. The pattern and colour of wing vary with the temprature and humidity.

Life Cycle

The mated female lay her eggs on tender leaves of the teak plant singly. A single female can lay about 250 eggs on the under side of the leaves. The eggs hatch within few days. The newly emerged larvae feed on tender leaves of teak. The larvae feed on entire green content of the leaf leaving veinlet network intact, thereby riddling the leaf. All instars can feed on young and older leaves of the plant. There are fire instars in the larval form. The last instar larvae pupate along with fallen leaves (green/dry). The pest hibernates from January to march. The life cycle is completed within one month.

Nature of Damage

Caterpillars feed voraceously on the tender and older leaves and skeletonize the plant affecting the growth.

Host Plants

Teak, some species of *Callicarpa*

Fig. 2.7 : *Eutectona machaeralis*

Control Measures

1) As above pest except encouragement of following biocontrol agents.

Biocontrol Agents

A) Parasitoids :

I) *Trichoqramma minutum* attack eggs (Trichoqrammatidae : Hymenoptera)

II) *Cedria paradoxa* (Braconidae : Hymenoptera) attack larvae.

III) *Trichoqramma pickel*
IV) *T. brasilensis*
V) *T. evanscence*

} attack eggs of the pest.

11) The hepialid caterpillar

Sahyadrassus malbaricus (Moore)

(Hepialidae - Lepidoptera)

Distribution

Kerala, Tamil Nadu, Maharashtra, Karnataka, Peninsular India.

Marks of Identification

The wing expanse of the moth is about 11 cm and body length is 5.5 cm. The moth is greyish brown. Eggs are creamy coloured. The full grown larva measures about 6.0 to 10.0 cm in legth and yellowish white and rarely black.

Life Cycle

Moths appear in may and lay eggs immediately. Eggs are laid on ground vegetaion. After the hatching of eggs, the newly emerged caterpillars develop on ground vegetation and then they migrate to saplings. On the saplings they prepare a long cylindrical tunnel along the pith. The tunnel may extend into tap root system. The tunnel is opened outside the plant by a mouth situated about 30-40 cm above ground. The larva feed on the callus tissue that develop around the tunnel mouth. Larva feeds at night under cover of particle-mat-cover, sometimes bark is browshed in a ring and saplings beak from this point. Full grown larva pupates in tunnel. The pupal period is one month. Life cycle is completed in one year.

Nature of Damage

Caterpillars are only destructive. They tunnel the stem even upto tap root system and affect the growth of plant. The larvae feed on callus tissue.

Host Plants

Eucalyptus, *Albizzia, Gmelina,* etc. This is polyphagus pest which feed on more than 40 host plants belonging to 22 families including woody shrubs and tree saplings.

Control Measures

I) Removal of ground vegetation and weeds check the breeding of larvae up on them.

II) Collection and destructive of all pest stages.

III) *Metarhizium anisopliae* and *Beauveria bassiana* are good microbial control agents of larvae of pest from fungi category. The above fungi cause mortalies in matured larvae of the pest.

Chemical Control

Spraying crop with 0.5% quinolphos is effective against above all lepidopterous pests.

3

PESTS OF DEODAR (*CEDRUS DEODARA*)

Sr. No.	Common Name	Scientific Name	Family	Order
1.	The melolonthid beetle	*Melolontha* sp.	Scarabaeidae	Coleoptera
2.	The Buprestid beetle	*Sphenoptera aterrima*	Buprestidae	Coleoptera
3.	The Buprestid beetle	*S. lafertii*	Buprestidae	Coleoptera
4.	The tenebrionid beetle	*Camarimena rugosistriatus*	Tenebrionidae	Coleoptera
5.	The cerambycid beetle	*Teledapus dorcadioides*	Cerambycidae	Coleoptera
6.	The cerambycid beetle	*Tetropium oreinum*	Cerambycidae	Coleoptera
7.	The deodar longhorn bast eater	*Trinophyllum cribratum*	Cerambycidae	Coleoptera
8.	The deodar weevil	*Brachyxystus subsignatus*	Curculionidae	Coleoptera
9.	The curculionid	*Rhycholus himalayensis*	Curculionidae	Coleoptera
10.	The blue spine scolytid	*Polygraphus major*	Scolytidae	Coleoptera
11.	The scolytid beetle	*P. aterrimus*	Scolytidae	Coleoptera
12.	The deodar branchlet	*Cryphalus deodara*	Scolytidae	Coleoptera

contd....

Sr. No.	Common Name	Scientific Name	Family	Order
13.	The Himalayan scolytid beetle	*C. himalayensis*	Scolytidae	Coleoptera
14.	The scolytid	*Tomicus stebbingi*	Scolytidae	Coleoptera
15.	The scolytid	*Pityogenes coniferae*	Scolytidae	Coleoptera
16.	The scolytus beetle	*Scolytus major*	Scolytidae	Coleoptera
17.	The scolytus beetle	*S. minor*	Scolytidae	Coleoptera
18.	The scolytus beetle	*S. deodara*	Scolytidae	Coleoptera
19.	The deodar wood borer	*Crossotarsus conififerae*	Scolytidae	Coleoptera
20.	The deodar defoliator	*Ectropis deodarae*	Geometridae	Lepidoptera
21.	The geometrid caterpillar	*Geometrina* Sp.	Geometridae	Lepidoptera
22.	The seed borer	*Phycita abietella*	Phycitidae	Lepidoptera
23.	The seed borer	*Euzophera cedrella*	Phycitidae	Lepidoptera
24.	The short horn cerambycid	*Trinophyllum* Sp.	Cerambycidae	Coleoptera

I) The deodar buprestid beetle

Sphenoptera aterrima Kerremams

Distribution

Western Himalya, Shimla, Bashhr, etc.

Marks of Identification

The beetle is 11.00 mm long and 4.00 mm broad. Antennal pores concentrated in a terminal pit. Scutellum is wide accumilating from behind. It has triangular clypeus of large size. The head of grub is yellow and small. Prothoracic segment is greatly enlarged in young grub. Body segments of grub are narrower towards posterior side in abdomenal region. Full grown grub is flat, elongate, narrow and yellowish in body colour. Pupa is also yellowish and elongate ovate in shape and exarate type.

Life Cycle

Mated female lays whitish rounded eggs on the bark or inner surface of the bast. In severe infestation the eggs are laid on fresh green bast of young tree. Female oviposites in June month. The eggs hatch within one week. Newly emerged grubs start feeding in the bast by preparing irregular chambers. The grubs go deeper and grooves both, the outer bark and sap wood when grown in larger size and approaching maturity. The full grown grub bore down into the sapwood and prepare pupal chamber by eating the content of the tree.The full grown grub pupates in pupal chamber made by itself. Pupal period is 4 to 6 months depending on climatic conditions. Pupal period is 6 months. The winter is passed in pupal stage or in immatured stage of imago. Thus, only one generation is completed by this pest in single year.

Nature of Damage

Both, grubs and beetles are destructive to Deodar. Grubs feed on the bast by making irregular chambers. The grub can go deep and bore the sapwood in any direction. Full grow larva bores down into the sapwood at an angle and prepare pupal chamber by feeding upon sapwood. The chamber size is always larger than the beetle size. A typical characterized enterance tunnel is located below the bark. However, the newly formed adult beetles find their way outside the sapwood tunnel by tunnelling the bark. The beetle eat the bark and makes the enterance tunnel for leaving the infected wood. Whitish-yellow sapwood is again the characteristic of infested tree because wood peckers work on the bark for collecting the grubs for feeding purpose. The insect has irregular feeding patterns. It can also girdle the tree and thus cause more rapid damage by killing the tree. The pest is destructive to the tree at all stages.

Host Plants

Deodar *Cedrus deodara*. It is specific pest of deodar.

Control Measures

(i) Infested trees should be burnt along with pest stages.

(ii) Taking off infested bark and its burning is also one of the effective preventive control measure.

(iii) The pest is internal feeder hence difficult to control. Therefore, encouraging natural enemies of the pest is important strategy of pest control. Specially birds like wood peckers which largely subsists upon pest species. Encouragement to following wood peckers are advised.

Biological control

(i) The pied wood pecker - *Dendrocopus himalayensis*

(ii) The scaly bellied green wood pecker - *Gecinus squamatus.* Both above species are good biocontrol agent of the pest.

(iii) Encouragement of parasitoids :

Hymenopterous parasitoids

Ephialtes viridipennis Morley (Ichneumonidae). The above parasitoid attack grubs of the pest. Therefore, its mass rearing and periodic release is advised.

Chemical control

Treating the crop with insecticides :

Dusting 5 % aldrin or 10 % BHC.

2) The deodar tenebrionid

Camarimena rugosistriatus Blair

Distribution

North west Himalaya

Marks of Identification

The beetle measures about 18.00 mm in body length. It is characterized by having short antennae and small head and prothorax. A thickset beetle has dull coppery brown colour, flat and depressed head. Elytra of the beetle is stout with straight base and strongly striate-punctate. Its legs are densely punctate.

Life Cycle

Life cycle of this beetle is not fully studied.

Nature of Damage

The beetles cause damage to deodar flowers by feeding upon

them. The beetles are associated with male flowers of deodar. They destroy petals of the young flowers by feeding upon them.

Host Plants

Deodar *Cedrus deodara.*

Control Measures

(i) Collection and destruction infested flowers along with pest.

3) The deodar cerambycid beetle

Teledapus dorcadioides Pas.

Distribution

Western Himalaya

Marks of Identification

The elongate beetle shows 8-10 segmented antenna which is about equal to the length of body. The beetle measures about 14 to 20 mm in body length and 3 to 4.5 in width. Head is larger and wider than thorax. In both sexes wings are absent. However, beetle has reddish brown to dark brown body colouration. Prothorax is with little protuberence medianly. Its scutellum shows dense grayish tawny pubescence. Femora is thickened slightly from the base. The first tarsal segment is longer than others in males while in females first segment is equal to other two adjoining segments.

Life Cycle

Eggs are laid on deodar bark in May/June months. After hatching the eggs within one week, the newly emerged grubs bore into the bark, bast and outer sapwood. The adult beetles emerge from the deodar tree in May or early June. Thus, only one generation is completed in a year.

Hosts

Deodar *Cedrus deodara,* Spruce *Picea morinda.*

Nature of Damage

The grubs bore into the bark, bast and sapwood thus affect the quality of wood.

Control measures

(i) Collection and destruction pest from infested trees and other host plants.

(ii) Pesticidal control as above pest.

4) The short horn cerambycid

Tetropium oreinum Gahan

Distribution

JK, UP, Northwest Himalaya, Western Ghats, etc.

Marks of Identification

The black or brownish black beetle measures from 9 mm to 14 mm in body length and 2.5 mm to 3.5 mm in width. Antennae are shorter than its body. Head is with tawny setae, prothorax is sparsely setose. The legs are some what glossy. Elytra is densely punctate and some what glossy near the base. The femora is fusiform and laterally compressed.

Life Cycle

Mated female oviposites in craks and crevices in the bark of green deodar trees which are standing sickly or newly felled. The eggs hatch within one week or so. The newly emerged grubs start feeding on the fresh sappy bast layer and then start grooving the sapwood. The grub works in zigzag manner in the long axis of the tree by preparing tunnel and feeding upon the wood content of the tree. The larva enjoy the life of totally internal feeder for about 5 to 6 months. The full grown larva pupate in a typically made pupal chamber. The pupal chamber is longish and typically narrow. The pupal chamber is prepared at the depth of half an inch in sapwood by full grown larva feeding upon the wood content of the tree. The pupal stage lasts for 6 months. When pupa moult into adult, the adult prepare outlet tunnel in the bark by feeding upon the bark and finally emerge from the enterance tunnel. The beetles mate in June or July and start egg laying on the bark of deodar trees. One generation of the pest is confirmed in a single year by some workers in India. However, some workers belive that there might be two generations in a year since, beetles are seen in summer and early autumn.

Host Plants

Deodar *Cedrus deodara.*

Nature of Damage

Both, grubs and beetles cause the damage to deodar tree. The grub feeds upon bark and sapwood and prepare tunnel or galleries. The galleries are long broad, shallow at the stage of larval form but it prepares a typical long and narrow pupal chamber for pupation in sap wood by feeding on woody content and thus damaging sapwood. The beetles cause damage while emerging from the pupal chamber by eating and preparing outlet tunnel for their emergence.

Control Measures

(i) The grubs are parasitized by hymenopterous parasitoids. They should by exposed to grubs of this pest for parasitization and killing by parasitoids.

(ii) Other control measures are similar to above pest.

(iii) Use of iron hooks for collecting and killing the pest stages from tunnels.

(iv) Petroleum/Chloroform/Kerosine soaked cotton balls filling in the infested tunnel and plastering with mud willl kill the pest inside the tunnel.

5) Deodar longcorn bast eater

Trinophyllum cribratum Bates

Trinophyllum sp. (Fig. 3.1)

Distribution

Western Himalaya, Northern Himalaya, J.K., Assam, etc.

Marks of Identification

The beetle measures from 11-13 mm in body length and 3.5 to 4.00 mm in width. They are chestnut brown coloured and with short fulvorus brown hairs. Head and basal segments of antennae finely rugulose punctate. Antennae shorter than body length. Eyes and antennal bases are typically rounded. Prothorax is semi-circular and shows two small circular black spots and feeble sinnate groove. Elytra closely and strongly punctured. Legs reddish brown. Abdomen typically narrowed towards posteriorly, abdominal first segment is longest. Grubs are whitish, prothorax is brownish, the full grown grub is with brown head and powerful black mandibles. The full grown grub measures about 28.0 mm in body length. Pupa is elongate and whitish yellow.

Life Cycle

Mated females lay their eggs on felled green trees of Deodar. The whitish rounded eggs are laid in crevices in the bark of tree singly. The eggs may also be laid on the northern or shady sides of standing trees. However, they are not laid on dry trees. The pest oviposits on large scale when sudden storm leads large number of wind falls in the forest.

The eggs hatch within one week or two. The newly hatched grubs immediatly start boring into the camibium layer by feeding upon the camibium content. The grub also grooves bast and sapwood by feeding upon them. The grub feed in the galleries of bast and sapwood until the first wave of cold of the winter. Later, pest goes into hibernating stage. Some individuals continued with boring habit but, majority of them go into hibernating stage. The grubs recommence feeding in April and in May, they commence with tunnelling into sapwood. The tunnel is angular or straight in section. The larva also prepare on elliptical elongate chamber parallel to the long axis of the tree. The full grown grub pupates in a pupal chamber. Larval period is 6 to 8 months and pupal period is 1 month to 6 weeks. It seems that pest completes only one generation during a year. The winter is passed in grub stage.

Nature of Damage

This is a serious pest of Deodar trees. The pest attacks only green trees. The larva bore into the bark, bast, cambium, sapwood; it make chambers/tunnels in the deodar wood. Due to tunnelling in cambium the vitality of the tree is adversely affected.

Host Plants

Deodar *Cedrus deodara*

Control Measures

(i) As per other cerambycid beetles.

Fig. 3.1 : *Trinophyllum* sp.

(ii) Collection and destruction of infested plant parts along with pest stages.

(iii) Collection and destruction of beetles at the time of oviposition and mating.

6) The deodar branchlet scolitid

Cryphalus (Cosmoderes) deodara Stebbing

Distribution

North West Himalaya.

Marks of Identification

The beetle is very small measuring about 1.8 mm in body length. It is shining dark brown to black in colour. The oblong beetle is shining and covered with whitish-yellow hairs. The head is concealed by thorax, prothorax not broader than long. Legs are rufous brown. Elytra is about twice the length of prothorax. Parellel sided but, declivous at apex. Its under surface and head is black. Larva is curved, minute and whitish.

Life Cycle

The beetles appears in June on deodar tree for laying their eggs. The female beetle lay her eggs near the girdle on the branchlet of deodar. The female girdle the branchlet near the base on the twig. On small twig, only one egg is laid but on large twig, several eggs are deposited by the female. The mother beetles, emerges in spring and girdles the branchlets. The eggs hatch with in few days. The newly emerged grubs mine up the twigs. The grub mines all round in the outer wood of the twig. A wood powder is found crumbling through such mines. The full grown grub pupates in the pupal gallery and emerge as an adult beetle from branchlet by short gallery which is made at right anlge to the long axis of deodar. Number of generations completed by this pest are not known.

Nature of Damage

The damage to deodar by this pest is of 4 kinds.

1. Girdling the branchlets at several places.
2. Tunnelling the branchlets.
3. Preparing horizontal galleries and exit holes.
4. The twig galleries are found running up and down the wood.

Host Plants

Deodar *Cedrus deodara*

Control Measures

1. Collection and destruction of beetles when they appear on the crop.
2. At middle of May the deodar twigs are fully infected, such twigs are to be cut down with pest stages and disposed off along with pest stages.
3. Treating crop with following Insecticides.

DUSAN (Denthoate) or Fenitrothion or 1-2 % aldrin 40 % EC in water.

7) The himalayan scolytid

Cryphalus himalayensis Stebbing

Distribution

Shimla, North east Himalaya, etc.

Marks of Identification

The small sized beetle measures about 1.6 mm in body length. It is ferruginous or blackish and shining and oblong. Legs and antennae shaft yellowish brown while, antennal club and tarsi are yellow. Prothorax slightly wider than long. Elytra is more than half as long. Under surface of the beetle is with fine yellowish pubscence. The grub is whitish and legless creature.

Life Cycle

The male beetle prepare mating chamber in the cambium layer by feeding upon it. The male select spot to work at the junction of two branches, from where it bore horizontally through the bark to cambium layer and prepare mating chamber. At least three females enter into the mating chamber prepared by male beetle for mating with single male. After mating females start working with mating chamber. She enlarges the mating chamber by feeding on the bast and sapwood and lay eggs on the enlarged portion of chamber in June and further in another generation in November. The eggs hatch within 2 to 4 days. Newly emerged grubs thus, located in the chamber near the living female. The grubs then bore from mating chamber either upwardly or downwardly in the stem. Their galleries thus formed are blocked with wood dust and excreta of larva and beetle. The full grown grub hollowout a longish chamber in the sap wood and pupate inside the chamber only. Two generations are completed in a year.

Nature of Damage (Fig. 3.2)

Damage is done by both, grubs and beetles by making galleries into the bark, bast and sapwood and also by making mating chambers by males and females.

Control Measures

(i) Collection and destruction of beetles.

(ii) Collection and destruction of infected plant parts along with pest stages.

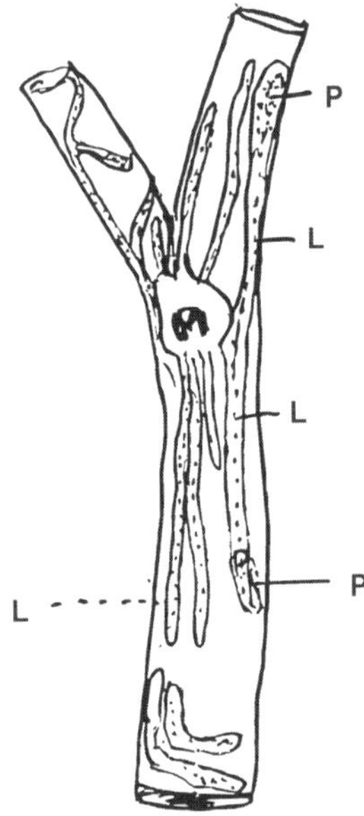

Fig. 3.2 : ***Cryphalus himalayensis*** **: Damage to deodar wood : P – Pupal chamber; L – Larval chamber; M – Pairing chamber**

(iii) Treating crop with following insecticides. 1-2 % aldrin 40% EC in water or DUSAN or fenitrothion or Endosulphan.

(iv) Encouragement of parasitoid, Ichneumon fly against grubs of the pest.

8) The pityogenes beetle

Pityogenes coniferae

Distribution

North West Himalaya, Baluchistan, etc.

Marks of Identification

The subelongate black beetles are measuring about 2 - 2.75 mm in body length. In male head is black while antennae are brownish. Elytra is shining and orange brown in males with lateral and basal portion black. Thorax is shorter and broader than in female. In females elytra is dark red brown or black. Grubs are whitish, legless, minute and curved. Pupa is whitish. Females are longer than males in body length.

Life Cycle

Mated females lay their eggs in egg gallery or egg chamber in April by first generation. The male prepare mating chamber or pairing chamber in sapwood by feeding on it. Later, about 5 females (virgin) find their way into the mating chamber one after another for mating purpose. The first entered female help male for enlargement of the mating chamber. The subsequent females either help in enlarging the original mating chamber or construct new chamber connecting the males chamber (Ist mating chamber). A single male can fertilize 5 to 6 females in the chamber. The fertilized females then immediately starts working for egg galleries which curves in one direction only from right to left. The five mated females work individually for ovipositon gallery and the galleries thus, have stellate appearence. The female prepare small notches in the margin of gallery and fill the egg in each. About 10 - 12 eggs are laid in each oviposition chamber/gallery. The eggs hatch within one week or so and newly emerged grubs start boring away from the mother gallery by feeding on bast layer mainly. The grubs prepare irregular curved galleries. The length of gallery ranges from 10 mm to 20 mm. The pest however, do not bore the sapwood. But, fullgrown grub finally prepare a gallery in the sapwood with its strong mandibles and pupate in it. Pupa is transformed into adult beetle. Tha matured beetle find its way out by entrance tunnel from pupal chamber. Thus, the beetle leave infested tree. The pest completes four generations in a single year from June, July, Sept, October or November. However, winter is passed in the larval stage.

Nature of Damage

Due to the making of numerous galleries and chambers by the pest the small branches and twigs start dying. The beetles prepare mating and ovipostion chambers. From central chamber (mating) 4 to 6 galleries are worked by individual females. Larval galleries are more or less right angle to the egg gallery. The larva finally, also bore into sapwood for pupal chamber. However, larva mostly feed on bast layer.

Host Plants

Blue pine *Pinus excelsa,* Spruce *Picea morinda,* Chilgoza pine *Pinus gerardiana,* etc.

Control Measures

I. Collection and destrucion of infectested branches and twigs along with pest stages.

II. Encouragement of following natural enemies.

a) Chalcid fly parasitizes grubs of this pest. Chalcid larva kill the grub by feeding upon it.

b) *Thanasimus* beetles feed on this pest.

c) Reduvid bug (Order : Hemiptera) feed upon the grubs of this pest by sucking the blood of grub with the help of strong rostrum.

III. Treating the crop with any of the following insecticides :

Dieldrin or endrin 0.2 % spray or As suggested for above pest.

9) The scolytid *scolytus* beetle

Scolytus major (Fig. 3.3)

Distribution

North West Himalaya

Marks of Identification

The black coloured beetle measures from 4.0 mm to 4.5 mm in body length. Its front of head is flat and antennae are rufos brown. Prothorax is constricted and impressed on anterior lateral margin, it is narrower anteriorly and broader posteriorly. It has elytra dark red, brown or black. Legs are brownish to black with tarsi rufosbrown.

Eggs are spherical and yellowish shining. The grub is whitish, small, curved and legs less. The pupa is whitish.

Life Cycle (Fig. 3.3a, b, c & d)

Mated females lay their eggs in egg galleries in egg-notches. A single mated female can lay about 70 to 85 eggs. In each side of the egg gallery, as much as 35 to 43 eggs are laid in notches. Egg hatching takes place within 2 days. The larva on hatching, bore into the bast and sapwood. Grubs prepare their galleries away from the egg gallery by feeding on the bast or sapwood. They work in a direction more or less right anlge and in an upward or downward direction. Thus, they have a special type of damage pattern (Fig. 3.4). The larval period is 4 weeks. Galleries prepared by first generation are strongest and longest one. Galleries may be slightly enlarged in large trees. Grubs bore down sapwood in poles, large trees and saplings. Finally the full grown grub prepare a pupal chamber in sapwood and pupate in it. The pupal stage lasts for about two weeks. Thus, the pest complete its life cycle, from egg to adult within 6 to 7 weeks. The pest complete many generations in a year. According to Stebbing (1977), pest completes 4 generations in single year. The first generation is completed from April to July, the second from June to August, the third from August to October and the fourth from Octrober to November. The pest pass the winter either in larval stage (in bast layer) or mature beetle (in thicker parts of old bark). Its fourth generation is partial. The beetle goes

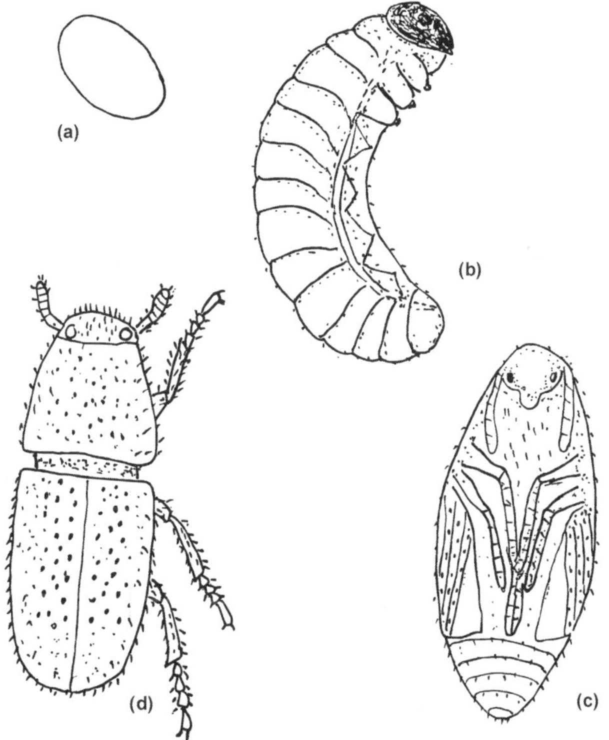

Fig. 3.3 : Life cycle : *Scolytis major*
a-egg, b-grub, c-pupa, d-adult (beette)

in hibernating stage for avoiding climatic danger, the cold of winter. Beetles bore the old bark for sheltring and protecting from cold and hibernates in it upto the first wave of warmth of the spring and then become active. Diapausing grubs also proceed with life cycle. They change to pupae in their galleries and within 7 to 10 days they become matured beetles and bore out of the tree and think for further life activities such as mating and oviposition. Beetles mate outside the tree or mating chamber. The mated female seek out the suitable place on the deodar plant for boring and for egg laying chamber. She commence with small shot-hole bore through the bark and the bast layer. She prepare a egg gallery maintly in the bast layer. The gallery may be of 2-3 inches in length in upward direction in a series of serpentine curves first to one side and second to the other. She also prepare egg notches in the gallery for deposition of eggs in them. She fix the egg in a notch with fine particles of wood dust. Excess wood dust particles which are whitish and yellowish are thrown outside the exit. The mother can reside in the gallery upto the egg hatching and then maturation of grubs and pupation, as parental care from parasitoids and predators. The mother finally blocks enterence tunnel with her body parts when she dies.

Nature of Damage (Fig. 3.4)

This is serious pest of deodar tree which tunnel the branches of deodar. The beetle can attack trees of all sizes from smallest sapling to largest tree. Upward flow of sap is affected due to tunnelling the plant and by that the vitality of the plant too. Beetles can cause damage to newly felled unbarked trees by making galleries. Beetles can attack green standing healthy trees. The pest is responsible for casualities upto 50 % at its maximum level and 15 to 20 % at minimum level. Beetles also damage standing sicky trees.

Host Plants

Deodar *Cedrus deodara*

Control Measures

I. Wind or snow-break fallen trees are collected and treated with pesticides.

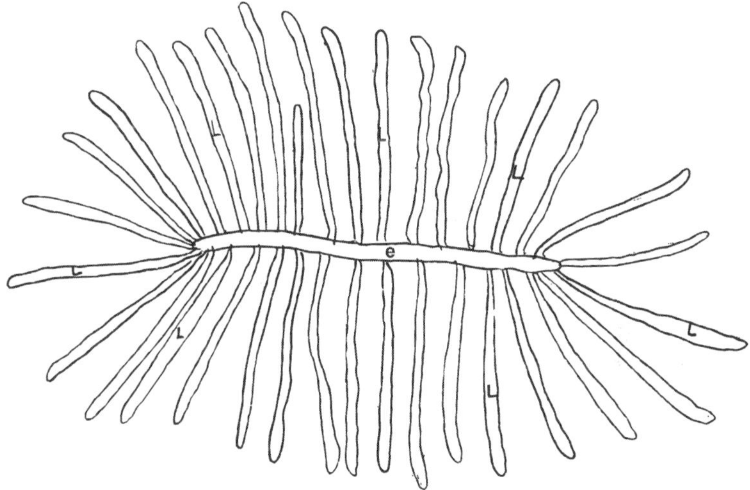

Fig. 3.4 : *Scolytus major* : Damage. e-egg gallery, L-Larval gallery.

II. Infested plants parts are collected and destroyed along with pest stages.

III. Inner face of bark is turned outward so as to expose pest stages to sun for killing grubs and pupae.

IV. Sapling and branches infested should be destroyed.

V. Green standing trees are used as trap crop for attracting beetles for oviposition and then collection and destruction.

VI. Newly felled green trees be used as trap trees. When the pest infestation peaks the trees should be barked and bark should be exposed to sun or it may be burnt.

VII. Encouragement to natural enemies :

a) The parasitic wasp, *Bracon* sp (Braconidae : Hymenoptera) is external parasitoid on the grubs of this pest. The parastic larva feeds on grubs and kill them.

b) A chalcidfly belongs to Order Hymnoptera (Family : Chalicidae) is also good biocontrol agent of the pest.

c) *Niponius canalicollis* Lewis (Hesterid beetle) also feeds on the pest in ovipositing chamber.

d) *Thanasimus himalayensis* (Coleoptera) predates upon this pest.

VIII. Chemical control :

Treating the crop with ecofreindly insecticides. Endosulphan or carbaryl at insecticidal concentration.

10) The major scolytid

Scolytus minor Stebbing

Distribution

North West Himalaya

Marks of Identification

This beetle resembles with *S. major* beetle but it is smaller than *S. major.* Its elytra is darker to black. It measures from 2.5 mm to 3.00 mm in body length. Its prothorax is constricted and rounded anteriorly. First abdominal segment is less rugose-punctate than *S. major.* The eggs, grubs and pupae are very similar to *S. major* but are small sized.

Life Cycle

Oviposition behaviour is similar to *S. major* except egg galleries are short in length. The life cycle is similar to *S. major* beetle in time and generations completed in a year.

Host Plants

Deodar *Cedrus deodara*

Nature of Damage

Beetles and grubs make numerous galleries in the bark, bast and sapwood.

Control Measures

I. As per *S. major*

II. Encouragement of natural enemies :

a) A braconid wasp (Hymenoptera : Braconidae) parasitizes grubs of this pest and cause mortalities.

b) Natural enemies of *S. major* also attacks this pest.

11) The scolytid deodora

Scolytus deodara (Fig. 3.5)

Distribution

Simla hills, Chamba, Bashahr, Jaunsar, etc.

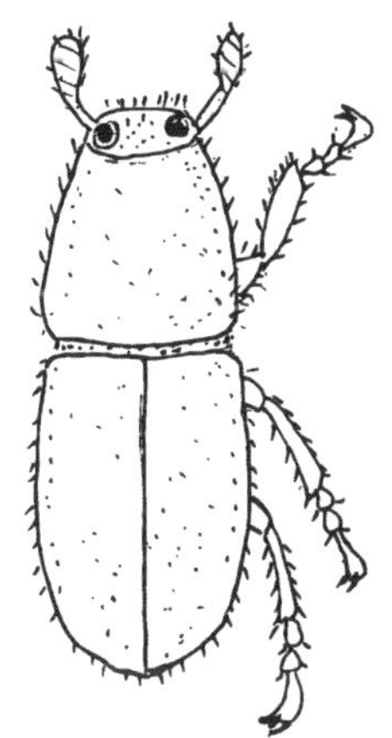

Fig. 3.5 : *Scolytus deodara*

Marks of Identification

The beetle measures about 3.5 mm in body legth, black, shining with elytra tinged with rufus brown. Prothorax longer than broad and tapers anteriorly. Flytra is black with yellow hairs. Antennae are yellowish brown and legs are rufus brown with tarsi yellow.

Life Cycle

Beetles appear in June. It lay eggs like other scolytids. Eggs are deposited on branches of tree. After hatching grub tunnels first into the cambium layer and then bore bast and sapwood in a horizontal manner. The groove completely girdle the stem. Full grown larva prepare gallery in sapwood and pupate in it. Many generations are possible in a year.

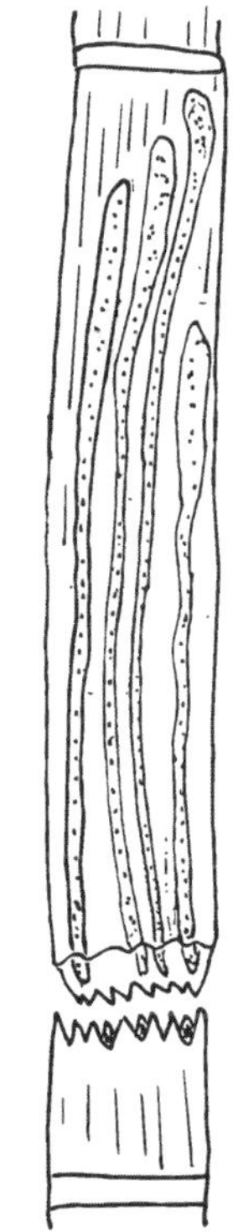

Fig. 3.6 : Larval galleries in sapwood

Nature of Damage (Fig. 3.6)

Beetles damage all plants but saplings most. Attacks to top shoots and spoil the shape of tree. Grub bore into the bast and sap wood and affect the growth of the tree.

Host Plants

Deodar *Cedrus deodara*

Control Measures

1) Collection & destruction of beetles along with infested plant parts.

2) Dusting the crop with 10 % BHC or 5 % Aldrin.

12) The deodar defoliator

Ectropis deodarae (Fig. 3.7)

Distribution

UP, MS, Kerala, Punjab, etc. It is recorded at higher altitudes, above 1500 m.

Marks of Identification

Moths are medium sized. Its forewings and hindwings are poorly developed and with numerous irregular black spots. Full grown caterpillar measures for 2 to 3 cm in body length. The pupa is brownish and obtect type. Eggs are rounded.

Fig. 3.7 : *Ectropis deodarae*

Life Cycle

Dipausing pupae get changed into moths in spring. Moths are weak fliers. After emergence from pupae, moths mate immediately. Mated females then start climbing on the deodar tree for egg laying on the needles of the tree. Eggs are laid in mass and they hatch within 2-5 days in spring. Thus, newly emerged larvae start feeding on the needles of the deodar. They moult for 5-6 times. Full grown larvae descend down on the ground for pupation. Larvae pupate on ground among the fallen needles or in soil under a layer of humas. The pest hibernates in pupal stage. However, the individual occurs in an epidemic form. It stays for at least 3 years once it occur on the crop. It may have its out break at 7-10 years interval.

Nature of Damage

Caterpillars are the only destructive stages of this pest. Caterpillars are most injurious to deodar tree since they feed on needles. In severe infestation, they skeletonize the plant.

Host Plants

Deodar *Cedrus deodara*

Control Measures

Preventive control :

I. Collection and destruction of egg masses.

II. Collection and destruction of caterpillars from deodar tree.

III. Digging field under the shade of the tree for exposing the pupae for natural mortality factors (biotic & abiotic).

IV. Clean clutivation for exposing pupae to biotic and abiotic mortality factors.

Biological control :

I. Use of parasitoids and predators.

II. Use of shrews in forest against pupae and matured larvae of this pest.

III. Encouragement of birds agianst larvae and pupae of the pest.

Chemical control :

Spray the crop with 0.03 % Azadirachtin (or) Dusting the crop with 5 % carbaryl 25 kg/ha.

12) The geometrid caterpillar

Geometrina sp.

Caterpillars of this pest are destructive to needle of the crop since they feed upon them. The pest has more or less similar pattern of the life cycle as recorded in *E. deodarae.*

Control Measures

As per above pest.

4

Pests of Shisham (*Dalbergia latifolia*)

Shisham *Dalbergia latifolia* is one of the important hard wood tree of Indian forest utilized for making furniture and other goods. It is widely grown in India and attacked by several insect pests. The list of which is given below.

Sr. No.	Common Name	Scientific Name	Family	Order
1.	The shisham bostrichid	*Sinoxylon anale*	Bostrichidae	Coleoptera
2.	The Shisham weevil	*Apoderus* sp.	Curculionidae	Coleoptera
3.	The Shisham beetle	*Serica* sp.	Scarabaeidae	Coleoptera
4.	The Apoderus weevil	*Apoderus sissu*	Curculionidae	Coleoptera
5.	The Shisham defoliator	*Plecoptera reflexa*	Noctuidae	Lepidoptera
6.	The Gram pod borer	*Helicoverpa armigera*	Noctuidae	Lepidoptera
7.	The Shisham weevil	*Adoretus caliginosus*	Scarabaeidae	Coleoptera

1) The shisham bostrichid
Sinoxylon anale Lesne

Distribution

In India, the pest is widely distributed in Punjab, Maharashtra, West Bengal, Belgaum, Chota Nagpur, Changa Manga, Dehradun, etc. Outside India, it has been reported from Myanmar.

Marks of Identification

Beetles are very small measuring about 4 mm to 5.5 mm in body length. Newly formed beetles are light yellow in colour but later they become dark and the outer paris hardens. Tubercles are on the front of the head. Antennae and palpi are testaceous. Antenna is with fan shaped club, second segment of the club is about 6 times as broad as long. Prothorax is very convex. Elytra is with two sharp teeth on apical delivity. This beetle is smaller than *S. crassum* in build & size. Grubs are whitish and curved, their anterior body segments are larger than others. Full grown grubs measure about 5 mm to 6.3 mm in body length. Pupae are whitish and beetle like in appearence.

Life Cycle

The pest has been reported on shisham in April 1901 from Changa Manga forest of India. It emerges from diapausing stage in different months i.e. March, April, May and January etc. in different parts of our country forests. Newly emerged and matured beetles tunnel into and oviposit on dying trees, dead trees, fresh cut logs and firewood billets and also old dry or rotting trees/material. Oviposition takes place in April or June or August or February as per the generations. For oviposition, beetle bores straight into the bark and then into sapwood and then prepare adeep pit like gallery where the sexes mate. This mating chamber may have one or two entrance tunnels. A male fertilizers at least two females as two egg tunnels are prepared. The eggs are laid at intervals in the egg tunnel. The newly emerged grub feed usualy in the long axis of tree or log. The larval chamber is blunt elliptical and free from wood dust and constructed parellel to long axis of the tree. The pest completes some over lapping generations in a year. It may hibernate either in larval or pupal stage or mature beetle. In Punjab the pest pass the winter either in larva or pupa or semi matured imago. Beetles hibernate in thicker bark of old dying or dead sissu trees. In Maharashtra, the pest completes 3½ to 4 generations in single year. In Tenasserin forest, the pest completes 4½ to 5 generations in a year. Beetles apear in January as a first generation.

Nature of Damage

The green and healthy trees are not attacked by this pest. The pest infests and kill sickly trees of shisham. It can attack wood at any where *viz*, main stem, branches or roots and thick knowty wood, the wood of main stem, etc. It also infest freshly felled trees of sissu.

Host Plants

It is polyphagus pest which affect the following trees in forest.

Phulahi *Acacia modesta, Terminalia tomentosa, Terminalia chebula,* Sissu *Dalbergia sisson, Dalbergia latifolia, Xylia dolabriformis, Prosopis spicigera, S. crassum,* Khair *Acacia catechu, Pterocarpus marsupium,* Bamboo *Dendrocalamus strictus.*

Control Measures

I. Collection and desrtruction of beetles when they emerge from the trees in various months.

II. Dead dying trees, rejected logs, large branches, decaying wood, etc are treated with any of the following pesticides.

Dusting 5% aldrin or chlordane.

Dusting 10% BHC.

Spraying 0.04% Fenitrothion

III. Natural Control

Encouragement of following natural enemies.

a) Predatory beetles :

I) *Teretriosoma stebbingi* - Feeds on grubs of this pest.

II) *T. cristatum* - Feeds on grubs of this pest.

III) *T. instrusum* - Feeds on grubs of this pest.

IV) *Bothrideres andrewesi* - attack larvae & pupae of the pest.

V) *Alindria* spp. predates the pest.

VI) *Malambia* sp. predates the pest.

2) The shisham weevil

Apoderus sp.

Distribution

Madras, Coimbatore, Western Ghats.

Marks of Identification

Weevils are not studied properly. However, their grubs are characterized by leg less appearence and their canary yellowish colouration. Its head is yellowish brown with brown mouth parts. First abdominal segment is enlarged and other segments taper posteriorly. The full grown larva measures about 4.6 mm in body length. Eggs are large, orange yellow coloured.

Life Cycle

Weevils appear in May/June and start egg laying on the tree in July. Eggs hatch at the end of July. Many weevils mate and oviposite in July or they may continue with the mating and oviposition and feeding. Weevils cause damage to leaves of *Dalbergia latifolia.* They cut the leaves and rollup the leaves. The above plant is badly attacked in July of the year. The full grown grub pupate on the ground among the fallen leaves and transformed into weevils.

Nature of Damage

Weevils are destructive to the crop since they cut the leaves and rollup the leaves. Thus, affect the growth and vigour of the tree.

Host Plants

Shisham *Dalbergia latifolia* and *D. sisso.*

Control Measures

I. Collection and destruction of leaves rolled up by the pest along with pest stages if any.

II. Encouragement of fungi and other natural enemies.

III. Chemical control

Dusting the crop with 5 % chlordane or aldrin (or) spaying the crop with 0.04 % diazinon or fenitrothion. or 0.03 % Methyl parathion or 0.02 % quinolphos or malathion 0.03 %.

3) The apoderus weevil

Apoderus sissu Marshall

Distribution

Punjab Changa Manga forest, Western Ghats, Kumaun forest, Pakistan.

Marks of Identification

Weevils are bright golden yellow. Males are smaller than females. Males measure about 1\6 of an inch while, female is 1/4 in inch. Edges of the head, thorax and elytra are black except, upper edges which are golden yellow. Antennae are yellow and stout and eyes are black. The elytra shows three black patches on their upper halves. Legs are bright golden yellow. Eggs are pale yellow and glistening.

Life Cycle

Adults (weevil) appear in May or June. They mate immediately and lay their eggs in June in certain parts of country and in May in Punjab. Eggs are laid singly and deposited to the left side of the midrib near the appex of the leaf or so. The leaf is cut down and is then folded along the midrib. The outer margin turn inward and the leaf is rolledup tightly. The rolledup mass hangs downwards. Many times the weevil cuts the leaf at about 1/8th or 1/6th of the length of leaf stalk. Due to the egg laying the leaf is cut down and later rolledup. Many times rolled portion of leaf is suspended in a very typical manner that it is just attached to very small edge of leaf portion of right side of the midrib. Eggs hatched into small grubs. Grubs feed upon stored food in the leaf galleries and later enter into the soil for pupation. Later, the pupa is transformed into adult weevil which appear in June for mating and oviposition on shisham leaves.

Nature of Damage

Damage is caused by weevils to leaves by oviposition and cutting and rolling the leaves and thus, affecting vigour of the crop. Branches of the tree in severe infestation are skeletonized i.e. they become leafless or the cut parts and rolled leaves drop down fastly on ground.

Host Plants

Sissu *Dalbergia sissoo*

Control Measures

I. Collection and destruction of leaf rolls along with eggs/grubs/beetles.

II. Clean cultivation will avoid pupation of pest in fallen leaves.

III. Plouging or digging soil for exposing pupae for natural mortality factors like biotic and abiotic.

IV. Treating the crop with any one of the following pesticides

a) Dusting 5 % chlordane or 10 % BHC.

b) Spraying 0.03 % methyl parathion or 0.02 % quinolphos.

4) The sissu scarabaeid

Adoretus caliginosus Burm

Distribution

UP (Gorakhpur), Western Ghats (MS).

Marks of Identification

The elongate beetle measures about 9.1 mm in body length and yellow with elytra lighter coloured than thorax. The head is black and prothorax is wider than long. Elytra is wider than thorax. The beetle is widest at apical third. The tibiae are with strong spines. Scutellum of the beetle is large.

Life Cycle

Beetles appear in May and mate in short time and then lay their eggs in soil. The beetle lay eggs singly. Newly hatched grubs feed on roots of sissu and some other plants. The full grown grub pupate in soil.

Nature of Damage

Beetles cause damage to foliage of sissu and grubs cause damage to roots from under ground by feeding upon leaves and roots respectively. Beetles have ability to skeletonize the tree entirely in severe infestation. Probably beetles come out at night from the soil for feeding on foliage part of the crop sissu. Beetles feed on foliage of both old trees and saplings.

Host Plants

Sissu *Dalbergia sissoo*

Control Measures

I. Collection and destruction of beetles from natural habitat.

II. Beetles get attracted towards light, therefore, light traps are used for collection of the beetles and their destruction.

III. Spray the crop with any of the following pesticides.

Carbaryl 0.15 % or 0.05 % Malathion or 0.05 % endosulphan.

5) The shisham defoliator

Plecoptera reflexa Guenee (Fig. 4.1)

Distribution

Punjab, Maharashtra, West Bengal, throughout north India, Andmans, etc.

Marks of Identification

The pest in its adult stage is moth. The moth is gray brown. Its head and collar are bright fulvous. Palpi slight and reaching vertex of head. The antennae of male are with long cilia and bristles. Thorax and abdomen smoothly scaled. Forewing with the apex nearly rectangular. It is with obliquely waved antemedially line. It also contains a large reniform spot with black centre and with a rufos spot on the costa. The forewing also shows an inwardly oblique waved post medial line and sinuous submarginal line. The marginal series of minute dark specks is again the characteristic of fore wing. The hind wing is fuscus brown with outer area slightly darker. Tibiae are slightly hairy and without spines. Males show fuscus area on costal margin of hind wing. A small tuft of hair is also located at the base of fore wing.

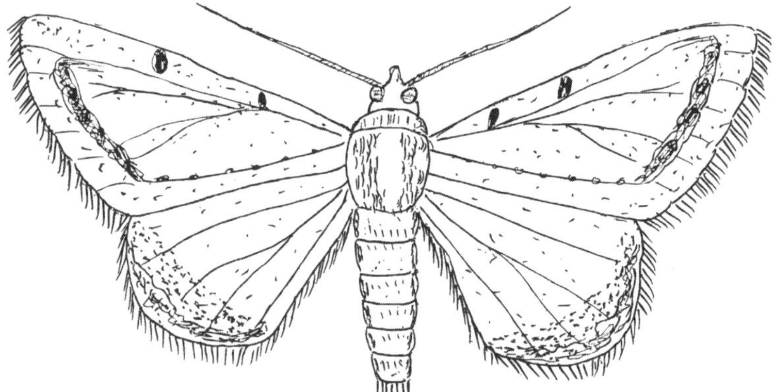

Fig. 4.1 : Plecoptera reflexa

Life Cycle

The mated female moth lay her eggs on tender leaves. Single female can lay about 400 eggs. Eggs are laid in clusters. They hatch within 4-6 days. On hatching the larvae feed on tender leaves of shisham *D. sissoo.* Larvae feed gregariously or solitarily on leaves of shisham. There are 5-6 instars in larval forms. The larval period lasts for 14-21 days and the pupal period for 5-12 days. Full grown larvae descend down on ground for pupation. They pupate among the fallen leaves on ground.

Nature of Damage

Shisham *Dalbergia sisoo.*

Control Measures

i. Collection and destruction of egg masses (clusters) and gregarious and solitary caterpillars from the leaves.
ii. Clean cultivation.
iii. Digging the soil near and under the shade and base of the tree for exposing pupae to natural mortality factors.
iv. Use of natural enemies like shrews and predatory birds.
v. Use of any one of the conventional pesticide given bellow.

Spray 0.035 % endosulphan or

spray 0.15 % carbaryl or

spray Azadirachtin 0.03 %

6. Gram pod borer

Helicoverpa (*Heliothis*) *armigera* (Hubn.)

Distribution

Cosmopolitan. In India it is widely distributed in most of the states.

Marks of Identification (Fig. 4.2a & b)

Moths are yellowish brown or straw coloured measuring about 20 mm in body length. A conspicous black spot is present at centre on fore wing and a dark speck or dark area near the outer margin. A black kidney shaped spot is present on ventral side of the fore wing. Hind wings are whitish or lighter in colour with a broad blackish patch along the outer and anal margin. The larva shows broken gray lines on the latera sides of back running parellel. Early instars are light orange or dark orange brown in colour. Colour change in older instars is the characteristic of this pest. In older instars the larval colour may be green or brown or black and may have the tinge of the colour of host plant.

Life Cycle

The fertilized female of *H. armigera* can lay about 740 eggs at its maximum extent. Eggs are laid singly on tender leaves of the shisham *D. latifolium*. They hatch within 2 to 4 days. Newly hatched larvae start feeding on tender leaves of shisham. There are five instars. All instars are defoliators, feed on leaves and skeletonize the plant in severe infestation. The larval period is 13-19 days depending upon climate. The full grown larva descend down on the ground for pupation and pupate in soil. The pupal period is 8-15 days. Thus, the life cycle is completed within 23 days and many generations are completed in a single year.

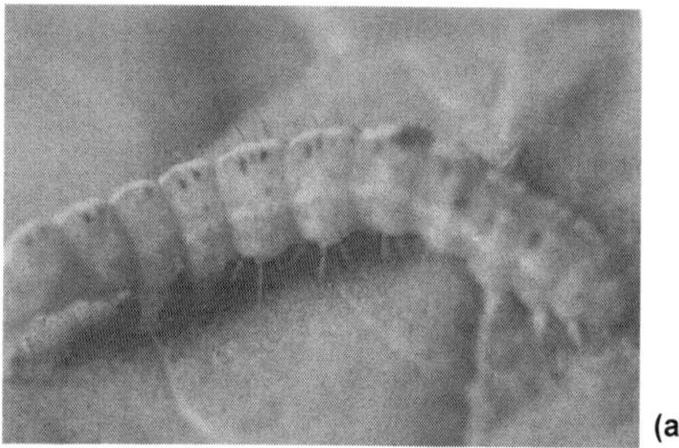

(a)

(b)

Fig. 4.2 : *Helicoverpa armigera.* (a) Larva (b) Adult (Moth)

Nature of Damage

Caterpillars feed voraceously on leaves and skeletonize the plant and thus, affect the growth and quality of plant. Severe damage to crop is noticed from July to August.

Host Plants

This is polyphagus pest. It attacks more than 180 host plants by damaging leaves, fruits, flowers and seeds. It is very bad pest of agricultural crops in India.

Control Measures

i. Collection and destruction of pest stages.

ii. Ploughing or digging the field for exposing pupae for natural mortality factors.

iii. Use of biocontrol agents

a. *Trichogramma* sp. 1,50,000/ha. (Egg parasitoid)

b. *Campoletis chloridae* attacks second instars of the pest.

c. *Apanteles ruficrus* attacks second instars of the pest.

iv. Chemical control

spray 0.1 % carbaryl or

spray 0.035 % endosulphan or

spray 0.03 % azadirachtin

v. Microbial control

a. Spray NPV (LE) 250 LE.

b. Apply DD-136 (Nematode parasites) 3×10^3 Juvenile/ml to the crop.

5

PEST OF INDIA RUBBER (*FICUS ELASTICA*)

Rubber is very important economic crop of forest in India. It is attacked by following insect pests.

Sr. No.	Common Name	Scientific Name	Family	Order
1.	The rubber passlid	*Leptaulax darjeelingi*	Passalidae	Coleoptera
2.	The rubber crysomelid	*Crioceris impressa*	Chysomelidae	Coleoptera
3.	The cerambicid beetle	*Xoanodera regularis*	Cerambycidae	Coleoptera
4.	The long horned beetle	*Batocera albofaciata*	Cerambycidae	Coleoptera
5.	The long horned beetle	*Batocera rubra*	Cerambycidae	Coleoptera

1. The rubber passalid

Leptaulax darjeelingi Knw.

Distribution

It is widely distributed in Assam, Darrang, Gopalpara and Munghu, etc.

Marks of Identification

Beetle measures about 20 to 28 mm in body length. The beetle is black shining with its flat head and numerous irregular tubercles. The mandible shows two teeth. Prothorax is with median longitudinal depressed line. Its elytra is striate punctate and blunty constricted apically. Under surface of the beetle is brownish.

Life Cycle

Females oviposite on rubber tree in the wood beneath the bark. Grubs bore into the soft decaying wood for feeding pupose. The full grown grub pupates in a pupal chamber prepared in wood. Adults thus formed leave the tree for mating and oviposition on India rubber tree.

Nature of Damage

Grubs damage wood by boring and beetles by ovipositing on the trees.

Host Plants

India rubber *Ficus elastica,* Semul *Bombax malbaricum,* Sal *Shorea robusta,* etc.

Control Measures

i. Beetles when appear on the crop they should be collected and destroyed by dipping them into keronized water.

ii. The rubber trees are treated with following insecticides.

a. Dusting 5 % aldrin/chlordane or

b. Dusting 10 % BHC.

2. The rubber crysomelid

Crioceris impressa Fabr.

Distribution

It is widely distributed in India. Out side India, it has been reported from China, Shri Lanka, Phillippines etc.

Marks of Identification

The broad robust beetle measures about 7 mm to 10 mm in body length. It is fulvous or piceous with silvery spots. It has bluish or blakish thorax. The elytra is fulvous. Antennae are much shorter and variable in colour. Prothorax and head are almost equal in breadth and more than twice the width of abdomen.

Life Cycle

Its life cycle is not fully understood. Sexes mate in April. Adults (beetles) cause damage to crop by feeding on the epidermis of leaves of India rubber tree.

Nature of Damage

Beetles cause damage by feeding upon leaves of India rubber plant. It cause damage to nursery plants.

Host plants

India rubber *Ficus elastica*

Control Measures

As like the passalid beetle

3. The Rubber Cerambycid Beetle

Xoanodera regularis Gahan

Distribution

Assam, North India, Myanmar, etc

Marks of Identification

The beetle measures about 17 mm to 20 mm in body length and 5 mm to 6 mm in breath. It is dark brownish to black in colour. Its elytra is with white-yellowish pubescence. Its prothorax is with longitudinal ridges of yellowish-white pubescence. Legs are with greyish white pubescence. Antennae are slightly longer than the body.

Due to light marking on elytra, the beetle is easy to identify. The beetle has typical dark patch on elytra on midlateral area.

Life Cycle

In April beetle appears from diapousing stage. The female beetle oviposite on the bark of the tree. Grubs bore and prepare larval gallery inside the tree between bast and sapwood. The grub pupates inside the gallery and transformed into adult beetle and escapes from the tree.

Nature of Damage

The damage is caused by grubs by feeding on bast and sapwood and costructing larval galleries.

Host Plants

India rubber *Ficus elasticus*

Control Measures

i. Collection and destruction of beetles in monsoon season.

ii. Cotton ball soaked with chloroform/petroleum/kerosine be placed in bore and sealed by plastering with mud.

4. The rubber longhorned beetle

Batocera albofaciata De Geer.

Distribution

Darrang, Assam, Western Ghats, etc.

Marks of Identification

The beetle is strongly built which measures about 30 mm to 36 mm in body length. Its antennae are longer than its body. The beetle has dull brownish purple or black colour. There are four distinct spots on elytra, two on right and two on left. Elytra is broadest at base with spine at shoulder. A pair of long sharp spines are located laterally on thorax. Grubs are flat, yellowish with orange prothorax and black head.

Life Cycle

Beetles emerge from diapausing stage in April and mate soon. Mated females lay their eggs on the bark of rubber tree or on the wounds of the rubber tree about the end of April or during May. Newly emerged grubs bore into the bark then reaches the bast by feeding on bark and bast. Grubs then bore deeper for gallaries for protecting them from cold winds. Grubs prepare more or less parellel galleries to the long axis of the tree. The gallery work is framed interior of large trees. When grubs full grown, they enlarges the size of gallery for pupation in heart wood and pupate in it. The larval gallery is packed with excreta of grub and wood refuse. Grubs pack the top and bottom of the pupating chamber. The pupal chamber is parellel to long axis of the plant. The larval period is 9 months and pupal period is 6 weeks to 2 months. Matured beetles come out of these pupal chambers and exit tunnels are being at right angle to the long axis. The tree infection/infestation can be detected by the exit orifice on the bark which are in good number. The pest completes only one generation in a year.

Batocera rubra Linn. *(Fig. 5.1)*

It has similar life cycle and damage as noticed in *B. albofaciata.*

Fig. 5.1 : *Batocera rubra*

Nature of Damage

Grubs are destructive to the rubber tree, they bore into the bark, bast and later heart wood by making galleries *viz*, larval galleries and pupal galleries. Beetles when emerge from pupal chamber, also find their way out by boring and thus affecting growth of the tree. In all cases, growth and vitality of the crop is adeversely affected.

Host Plants

India rubber *Ficus elastica.*

Control Measures

i. Badly affected trees should be cut out, cut up, and burnt along with pest stages.

ii. Collection and destruction of beetles in April will help to suppress future attack by the pest.

iii. The wounds on the rubber tree should be treated with conventional pesticides.

iv. Bores plugged with chloroform/petroleum/kerosine soaked cotton balls and plastering with mud will kill the pest inside the tunnel/bored portion.

6

Pests of Asan (*Terminalia tomentosa*)

Asan *Terminalia tomentosa* and Arjun *Terminalia arjuna* are very important plants of forest ecosystem. Both above plants have tremendous importance in Tasar sericulture. Tasar silk worm *Antheraea mylitta* is reared on above plants. Timber of these plants is used for making variety of items like cabinets, doors, window, furniture, cricket bats, etc. In India Asan is grown in United provinces, Central provinces, Seoni, Banjat valliy, Manda, Kowloon Island, Salween River, Siwaliks, Terai forests, Oudh, Wutgyi, Tenasserim, Thano, Dehradun, etc. Asan and Arjun plants are attacked by more than 25 insect pests. Important insect pests of Asan are listed below :

Sr. No.	Common Name	Scientific Name	Family	Order
1.	The cucujid	*Platycotylus inusitatus*	Cucujidae	Coleoptera
2.	The bostrichid	*Sinoxylon crassum*	Bostrichidae	Coleoptera
3.	The bostrichid	*Sinoxylon anale*	Bostrichidae	Coleoptera
4.	The bostrichid	*Psiloptera viridans*	Buprestidae	Coleoptera
5.	The buprestid	*Chrysobothris indica*	Buprestidae	Coleoptera
6.	The monommid	*Monomma brunneum*	Monommidae	Coleoptera
7.	The cerambycid	*Aeolesthes holosericea*	Cerambycidae	Coleoptera
8.	The cerambycid	*Xylotrechus* sp.	Cerambycidae	Coleoptera
9.	The brenthid	*Ceocephalus carus*	Brenthidae	Coleoptera

contd...

Sr. No.	Common Name	Scientific Name	Family	Order
10.	The scolytid	*Sphaerotrypes siwalikensis*	Scolytidae	Coleoptera
11.	The scolytid	*S. globulus*	Scolytidae	Coleoptera
12.	The scolytid	*Sphaerotrypes* sp.	Scolytidae	Coleoptera
13.	May-June beetle	*Anomala* sp.	Melolonthidae	Coleoptera
14.	The thrips	*Mallothrips indicus*	Thripidae	Thysanoptera
15.	Termites	*Microtermes* sp.	Termitidae	Isoptera
		Odontotermes sp.	Termitidae	Isoptera
		Trihervitermes sp.	Termitidae	Isoptera
16.	The shoot borer	*Sphenoptera koenbierensis*		
17.	The gall fly	*Phylloplecta hiruta*		
		Trioza fletcheri minor		
18.	Vapourer tussock moth	*Notolophus antiqua*	Lymantriidae	Lepidoptera

1. The sinoxylon beetle

Sinoxylon crassum Lesne. (Fig. 6.1)

Distribution

Dehradun, Punjab, Raipur, Mandla, Assam, Belgaum, Karnataka. It is also reported from Myanmar.

Mark of Identification

The beetle is elongate, oblong and stout which measures about 7 mm to 8.5 mm in body length and dark brown to black in body colour. Prothorax is very covex and rounded in front. Elytra is parellel sided, convex and apex is truncate. Head is vertical and hidden beneath the prothorax. Its antennae and legs are reddish brown. Grubs are whitish creatures.

Life Cycle

The female beetle lay her eggs on variety of trees mentioned under host plants on felled timber those which are not too dry. Beetles generally does not infesting completely dry timber. For oviposition purpose, the female tunnels into the wood either through bark on cut log or billet. Its end is enlarged into the mating chamber. Thus, he chamber has space in which two beetles (♂&♀) can move about. Male beetle enters in this chamber and fetilize the female present in the chamber. The mating chamber is then carried to the heart of the tree or heart wood. Its direction is changed several times. The male fetilizes more than one females. He pairs with the female more than a single time. Egg tunnels vary in length. The end of the tunnel is usually parellel to the long axis of the tree. Egg chamber is prepared within 4 to 8 days. Eggs laid in chamber hatch within 48 hours. Newly hatched grubs start feeding entirely on the wood, eating out tunnels in the long axis in an irregular manner. In old trees these tunnels are confined to sapwood region only. Full grown grubs pupate in a slightly enlarged chamber at the end of their galleries. The larval period lasts for 4 to 6 weeks, the larval period is extended upto 8 to 10 weeks in colder region of our country.

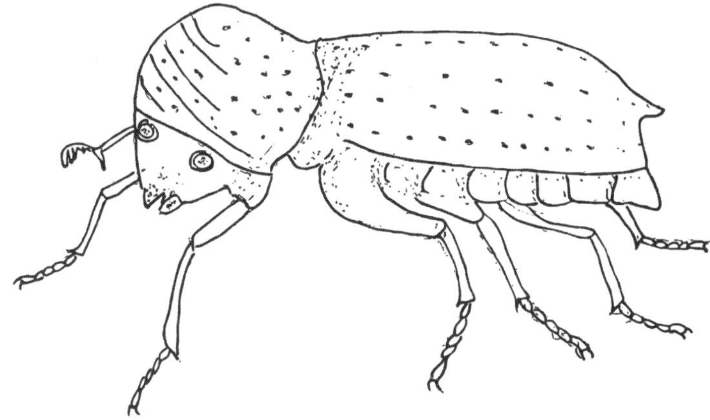

Fig. 6.1 : ***Sinoxylon crassum***

Nature of Damage

Damage is caused by females first to wood or sapwood of Terminalia by making egg tunnels and mating chamber and secondly, by grubs, by preparing tunnels in wood or sapwood or heart or the wood by feeding.

Host Plants

Asan *Terminalia tomentosa,* Khair *Acacia catechu,* Harra *Terminalia chebula,* Sal *Shorea robusta,* Sisso *Dalbergia sissoo,* Phulahi

Acacia modesta, Albizzia procera, Prosopis spicigera, Anogeissus latifolia, Pterocarpus marsupium etc.

Control Measures

i. Removal of felled trees from forest.
ii. No fresh cut-wood being allowed to lie.
iii. Reduce breeding places of the pest.
iv. Treating the tree with methyl parathion 0.03 % spray.
v. Use *Bracon* sp. (Hymenoptera : Broconidae) against the pest.
vi. Use following predators for controlling the pest *Teretriosoma* sp., *Teretrius* sp., *Alindria* spp., etc.

2. The bostrichid beetle

Sinoxylon anale Lesne (Fig. 6.2)

This pest is discussed under pests of sissu.

Fig. 6.2 : *Sinoxylon anale*

3. The buprestid

Psiloptera viridans Kerr.

Distribution

Karnataka, Mandla, Sylhet, etc.

Marks of Identification

Beetles are 14 to 23 mm long and 4.5 to 7.0 mm broad. They are oval, elongate, greenish golden. Head is flat. Elytra is scarcely wider than thorax at base. Legs are coppery and punctate. Larva is elongate, white and with yellow head.

Life Cycle

Beetles appear in February and start egg laying on tree of *Terminalia.* Grubs after hatching from eggs feed on woody content and or on sapwood. The larval gallery is long, irregular and winds about a different angles. The gallery prepared by grub is broad and found packed with dust and excreta of the pest species. Full grown larva pupates in a pupal chamber prepared by larva by eating parellel to the long axis of the tree. Pupal and larval chambers are found connected. The pest over winter in pupal stage in pupal chamber.

Nature of Damage

Females cause damage to stem of the tree by boring and making egg laying chambers and mating chambers in wood. Similarly, grubs bore into wood and make larval galleries and finally pupal chambers. Thus, affecting growth and vitality of the tree.

Host Plants

Sal *Shorea robusta,* Asan *Terminalia tomentosa,* Arjun *Terminalia arjuna,* etc.

Control Measures

i. Collection and destruction of infested plant parts along with pest stages.
ii. Other control measures as suggested for *S. crassum* beetle.

4. The crysobothrid beetle

Crysobothris indica (Fig. 6.3)

Distribution

Andaman Islands, Kowloon Island, etc.

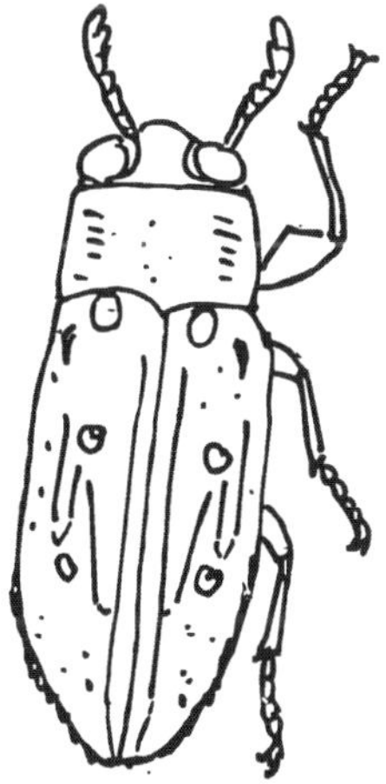

Fig. 6.3 : *Chrysobothris indica*

Marks of Identification

Beetle measures about 12.5 mm in body length and 5 mm in breadth and having dull coppery bronze body. Its head is with transverse carina between eyes. Prothorax sides are concave. Elytra is with 3 highly punctate coppery red or yellow despressions. Legs are coppery anterior femora is unidentate. Last abdominal segment is coppery and carinate down middle.

Life Cycle

Beetles emerge in December or January from the trees and females lay eggs either in crevices in bark or on green sicky *Terminalia* trees or on newly felled trees after mating. Incubation takes place within a week. Newly emerged grubs feed on the bast layer at first and then start boring into sapwood and remains in outer layers. It prepare winding and shallow gallery by feeding on wood content and enjoy internal feeding life. The gallery prepared by the larva is packed with wood dust and excreta of the grub. (Fig. 6.3a) Galleries are not unidirectional, they are more or less parellel to the long axis of the tree. Many times galleries are found towards right or left side. The larva increases in size quite fastly after first moult hence gallery breadth also found increased. At the end when larva is full grown, it prepares larger pupal chamber for pupation. This chamber is rather broad and oval in shape and free from wood excreta and dust and have the length of an inch and breadth of 1/4 inch. The full grown larva pupate in this chamber. Later, the pupa is transformed into the adult beetle in the same chamber. When adult is formed, the beetle stay in the same chamber, enjoy the free movements and then start eating the bark for preparing an exit for emergence from the tree. The pest can complete two generations in a year. Thus, beetles of first generation appear in December-January and of second generation in September-October.

Nature of Damage (Fig. 6.4)

Beetles cause damage to crop at the time when they want to come out from pupal chamber by boring and eating the bark of the trees. Grubs cause damage by preparing galleries and chambers to the wood of host plant.

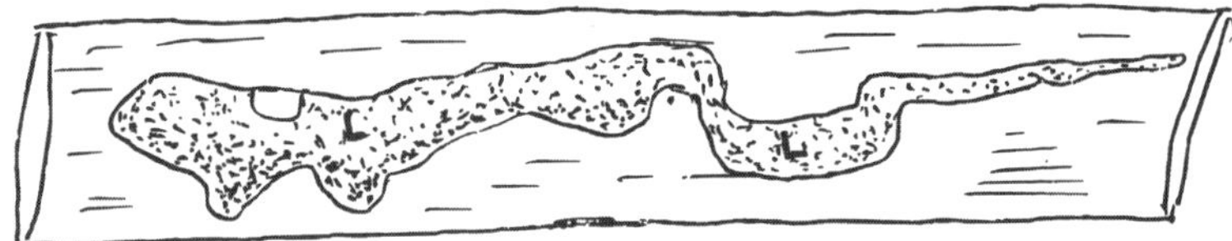

Fig. 6.4 : *Crysobothris indica* : damage to Asan tree. p-pupal chamber, L-larval gallery in bast & sapwood.

Host Plants

Asan *Terminala tomentosa* and several other forest trees.

Control Measures

i. Collection and destruction of beetles when they emerge in months, December-January and September-October each year.

ii. Removing newly cut trees and felling trees from forest.

iii. Treating the crop with following insecticides. DDVP 0.03 % spraying or dusting 10 % BHC.

5. The united provinces sal longicorn

5. *Aeolesthes holosericea* Fabr.

This pest is described under pests of Babul. Its life cycle on *T. tomentosa* and *A. arabica* is very similar.

6. The benthrid beetle

Coecephalus carus Walk.

Disrtibution

Dehradun, North India, Sri Lanka, etc.

Marks of Identification

Beetles are small sized measuring about 18 to 19 mm in body length and dark red and shining. Its rostrum is as long as prothorax. Legs are long with femora swollen anteriorly. Elytra is elongate, narrow and constricted at apex.

Life Cycle

Eggs are laid on stem. Grubs bore into the bark. Pupation takes place in the gallery. Adults emerge from the gallery and leave the tree.

Host Plants

Terminalia tomentosa

Control Measures

As above pest

7. The sal scolytid

Sphaerotrypes siwalikensis Steb.

Distribution

Siwaliks, Tarai forest, Oudh, etc.

Marks of Identification

Beetles measures about 2.75 mm to 3.5 mm in body length. They are black with reddish tinge on thorax and basal portion of elytra and antennae are yellow. Head is black. Legs are brownish with tibiae serrate on edge and tarsi with lighter colour. Under surface is black. The larva is white and curved, legless with brownish head. Pupa is white and spherical in shape and exarate type.

Life Cycle

Life cycle is similar as on sal tree. Hence given under pests of sal.

Nature of Damage

Beetles cause considerable damage to *Terminalia*. Beetles attack felled trees and green trees in forest. The beetle tunnels the bast for oviposition and the grub after hatching from the egg also groove the bast and sapwood for feeding purpose and later for preparing pupation chamber.

Host Plants

Sal *Shorea robusta, Anogeissus latifolia,* Asan *Terminalia tomentosa,* etc.

Control Measures

i. Felled trees should be removed from forest.

ii. Felled trees may be used as trap tree. The entire population of pest should be barked as soon as they are full of larvae.

iii. Use following parasitoid for killing pest.

a. *Bracon* sp. (Braconidae : Hymenoptera). This parasitoid cause mortalities in grub stage by parasitism.

iv. Use of following predator

Niponis andrewesi Lewis : Grubs of this beetle feed on the grubs of pest species and cause mortalities.

8. The scolytid beetle

Sphaerotrypes globulus Blandford

This pest is described under pests of Sal.

9. Termites

Microtermes sp.

Odontotermes sp.

Trinervitermes sp.

Distribution

Throughout India

Marks of Identification

Termites or white ants are social insects. There is polymorphism in them. They show caste system and division of labour. They digest cellulose.

Life Cycle

During swarmming winged reproductives emerge in large numbers. Within half an hour, wings of the individuals get sheded and termites settle down on the ground. The individuals select the mate and findout for nesting chamber/mating wherein the pair mate fequently and the female start egg laying in the chamber. The first generation is governed by both the individuals i.e. queen and king. Second generation onwards different castes perform their own functions. Many generations are completed in a year.

Nature of Damage

Termites construct their nest (termiteria) on the tree trunk and feed on woody content and bark resulting in retardation of growth and later death of the plant. Termites also damage seedings in nurseries.

Control Measures

i. Sprinkling water over termiteria (Nest of termites) will protect the crop.

ii. Treating the soil with 5 % Dust of Aldrin/chlordane.

iii. Treating the crop with Aldrex-30 EC or Dieldrex - 18 EC with 0.1 to 0.3 % spray or endrex - 20 EC at 0.20 % spray.

10. Vapourer tussock moth

Notolophus antiqua

Distribution

Western Ghats, Maharashtra, Karnataka, Kerala, TN, etc. It is world wide distribution except Neotropical region.

Marks of Identification

In male moth vein 9 arises from 10 and anastomosing with 8 to form an areole on forewing. Legs are heavily fringed with hairs. Palpi are short and heavily fringed with hairs. Female palpi and antennae are less hairy. Antennae are serrate type. Wings are aborted, scale like and covered with hairs.

Life Cycle

Eggs are laid in clusters on tender leaves of *T. tomentosa* and *T. arjuna* plants. Eggs are hatched within a week or so. The larvae feed on tender leaves and pass through 5 instars to full grown. Full grown larvae pupate either on the plant or on ground along with fallen leaves. Caterpillars feed voraceously on the leaves of *T. tomentosa* and *T. arjuna*.

Nature of Damage

Caterpillars feed on leaves voraceously and affect the growth of plant adversely reducing leaf yield of crop for sericulture bussiness. In severe infestation, the seedlings and young plants are completely consumed.

Host Plants

Asan *Terminalia tomentosa,* Arjun *Terminalia arjuna.*

Control Measures

i. Collection and destruction of eggs, larvae and pupae.
ii. Spraying the crop with Azadirachtin 0.03% or Endosulphan 0.05% or DDVP 0.03%.
iii. Clean cultivation.
iv. Biological control

Release *Trichogramma chlonis* 2,00000/ha at an internal of 3 days. *T. chilonis* (Trichogrammatidae - Hymenoptera) is egg parasitoid of the pest.

7

PESTS OF OAK (*QUERCUS* SPP.)

Different species of Oak (*Quercus*) are widely cultivated in Indian forests. Oak plants are very good source for rearing tasar silkworms (*Antheraea proyeli*) in India. Oak plants are used as timber and making various wooden articles used in day to day life of Indians. Hence Oak flora have tremendous importance in Indian forests. Important species of the genus *Quercus* grown in India refers to *Q. griffithi, Q. dilatata, Q. incana, Q. ilex, Q. glauca, Q. semicarpifolia, Q. himalayana, Q. serrata, Q. semiserrata,* etc. Out of which later seven species are well known food plants of temperate tasar silk worm (*A. proyeli*) and are grown in western and subhimalayan belt and eastern belt of Indian forests. *Q. griffithi* is widely grown in the forests of Nainital, Kotgahr, Bashahr, Kashmir, Punjab, Assam, Taunggyi and Kumaun oak forests. The Ban Oak *Q. incana* is widely grown in Chamba forests and North West Himalayas. While, Moru Oak *Q. dilatata* is grown in the forests of Jaunsar, Kilba, Almora, Nainital and North West Himalayas. Above mentioned species of quercus are broadly attacked by a very large number of insect pests. The most important amongst them are listed in the following table :

Sr. No.	Common Name	Scientific Name	Family	Order
1.	The stag beetle	*Lucanus lunifer*	Lucanidae	Coleoptera
2.	The Oak Timber longicorn	*Lophosternus hugelii*	Cerambycidae	Coleoptera
3.	The Oak cerambycid	*Paraphrus granulosus*	Cerambycidae	Coleoptera

contd...

Sr. No.	Common Name	Scientific Name	Family	Order
4.	The Oak cerambycid	*Massicus unicolor*	Cerambycidae	Coleoptera
5.	The mango stem borer	*Batocera titana*	Cerambycidae	Coleoptera
6.	The meges cerambycid	*Meges marmoratus*	Cerambycidae	Coleoptera
7.	The Ban Oak curculionid	*Apoderus incana*	Curculionidae	Coleoptera
8.	The Acorn weevil	*Calandra sculpturata*	Curculionidae	Coleoptera
9.	The Oak scolytid	*Chramesus globulus*	Scolytidae	Coleoptera
10.	The moru Oak scolytid	*Dryocoetes hewetti*		
11.	The Ban Oak platypodid	*Crossotarsus fairmairei*	Platypodidae	Coleoptera
12.	Oak Aphid	*Cervaphis quercus*	Aphididae	Hemiptera
13.	The Oak *Trabala* caterpillar	*Trabala vishnu*	Lymantriidae	Lepidoptera
14.	The Oak caterpillar	*Tortrix* sp.		Lepidoptera
15.	The Oak Tinea worm	*Tinea* sp.		Lepidoptera

1. The stag beetle

Lucanus lunifer Hope. (Fig. 7.1, 7.2)

Distribution

Himalayas

Marks of Identification

Black or greenish black beetle is large sized, measuring about 35 to 75 mm in body length. It has shining elytra of coppery brown colour. Vertex is shield like, front of head slopes downwards, the anterior edge with a median bifurcate prolongation. Mandibles are very special which shows two long horns like stags antlers hence the name stag beetle. It shows elbowed antennae which are quite long. Prothorax is wider than long, black coloured. Elytra conjointly rounded at apex. Legs are long with spiny tibiae. Its abdominal segments are brown. (Fig. 7.2) The males are larger than females (Fig. 7.1). Thorax and median bifurcated prolongations are longer than female. Its larva is corrugate and having 3 pairs of thoracic legs measuring about 50-85 mm in body length when full grown.

Life Cycle

Beetles appear in June, July and September. The mated female lay her eggs on dead trees specially in crevices of the bark or creeps under projecting flakes and on outer surface of sapwood. After hatching the eggs, the newly emerged grubs start boring into timber and full grown in tunnel. Its larval period is unknown. It is belived that larva can live for 3 to 4 years in interior of trunk tunnel. The grub bore into heart of the stem winding into various passages both up and down the trunk. The pest has short pupal period of one month to 6 weeks. This pest is most destructive to living trees since beetles and grubs destroy the oak timber.

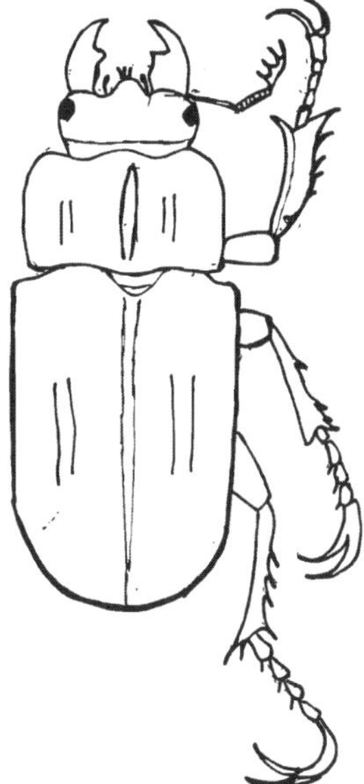

Fig. 7.1 : *Lucanus lunifer* (Female)

Fig. 7.2 : *Lucanus lunifer* (male)

Nature of Damage

Both grubs and beetles are destructive to oak timber by making tunnels and feeding upon it. This pest is most destructive to oak tree.

Host Plants

Moru oak *Quercus dilatata*, Ban Oak *Q. incana*

Control Measures

i. Collection and destruction of beetles when they appear in June, July and September.
ii. Removal of felled trees from forest.
iii. Spraying the crop with 0.05 % Nuavacron or dusting the crop with 10 % BHC or 5 % Aldrin.

2. The oak timber longicorn borer

Lophosternus hugelii Redtenb. (Fig. 7.3)

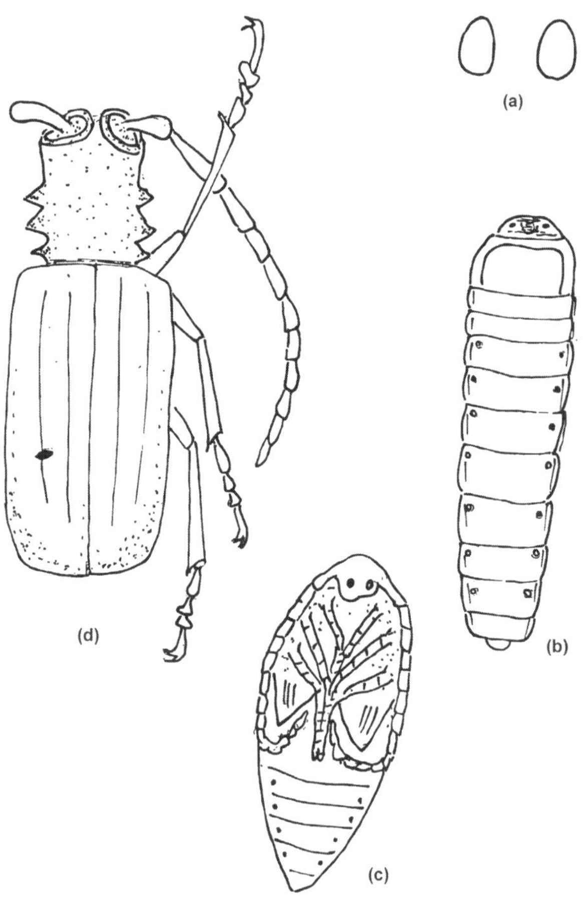

Fig. 7.3 : *Lophosternus hugelii* : Life cycle : (a) eggs, (b) grub (c) pupa (d) adult (Beehe)

Distribution

Assam, Punjab, Kashmir, Northwest Himalaya, Nainital, Mussoorie, etc.

Marks of Identification

Chest nut red coloured beetle measures from 30 to 52 mm in body length. Its head and prothorax are darker than elytra. Its eyes are large. Antennae are little shorter than body. First segment of antenna not reaching beyond the hind margin of the eye. Its third to fourth segments acutely produced at apex on the anterior side. In female the antennae hardly reach to middle of elytra and are more slender than in male. Its larva is yellowish white and with strong black mandibles. Pupa is yellowish white and exarate type.

Life Cycle (Fig. 7.3a, b, c, d)

Beetls appear on the wing in July and August and mate immediately and start egg laying immediately. Incubation takes place within few days. The oviposition takes place on the bark of the plant. After incubation newly emerged small grubs bore into bast and then sapwood and feed upon the tunnelling matter. Young sapwood are slightly bored by this pest. Grubs when fullfed tunnels down into the wood for pupation. The grub thus, prepare pupal chamber of larger diameter parellel to long axis of the tree.Excreta of wood dust are not located in pupal chamber. The adult ermerge out from pupal chamber by making tunnel for exit. Life cycle of this pest is completed in about more than one year.

Nature of Damage

Grubs destroy bast layer by feeding on it. The grub also cause damage by tunnellng into sapwood and heart wood resulting the death of the plant.

Host Plant

Ban Oak *Quercus incana,* Moru Oak *Q. dilatata.*

Control Measures

i. Collection and destruction of beetles in July and August when appear on wing and congregrate on oak tree.

ii. Collection and destruction of infested plant parts along with pest stages.

iii. Encouragement to natural enemies such as *Loranthus* sp. which cause mortalities in the pest.

3. The oak cerambycid beetle

Massicus unicolor Gahan. (Fig. 7.4, 7.5)

Distribution

Assam, Doherty, etc.

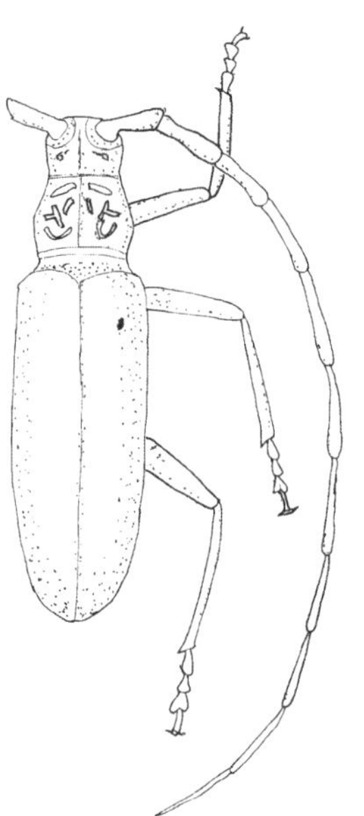

Fig. 7.4 : *Massicus unicolor* (Male)

Fig. 7.5 : *Massicus unicolor* (Male)

Marks of Identification

The beetle is stout and large sized measuring about 60 mm in body length and 16 mm in width. Females are smaller than males and have shorter antennae. Male is blackish brown covered with short pubescence of greyish tint. Antennal length is not twice the length of its body. Antennal first segment not reaching to the prothorax. Prothorax is some what rounded at sides, constricted a marked with transverse furrow at appex two furrows, near the base, the entire surface of prothorax is irregularly corrugated. At apex, the elytra is transomely truncate and its outer angles are obtuse. At the base of each antennac 'U' shaped marking is present. Eyes are large and black. Larva is ivory white, stout, corrugated and long while, pupa is yellowish and exarate type.

Life Cycle

Beetles appear in September and lay eggs in the same month on the oak tree. A mated female can lay about 45 to 48 eggs. After hatching the eggs, newly emerged grubs start boring the stem. The grub move apparently up and down by tunnelling the wood. The tunnels are elliptical in shape. The larva (Fig. 7.6a) make typical noice at the time of tunnelling the wood. Larval period is two years while pupal period is 6 weeks.

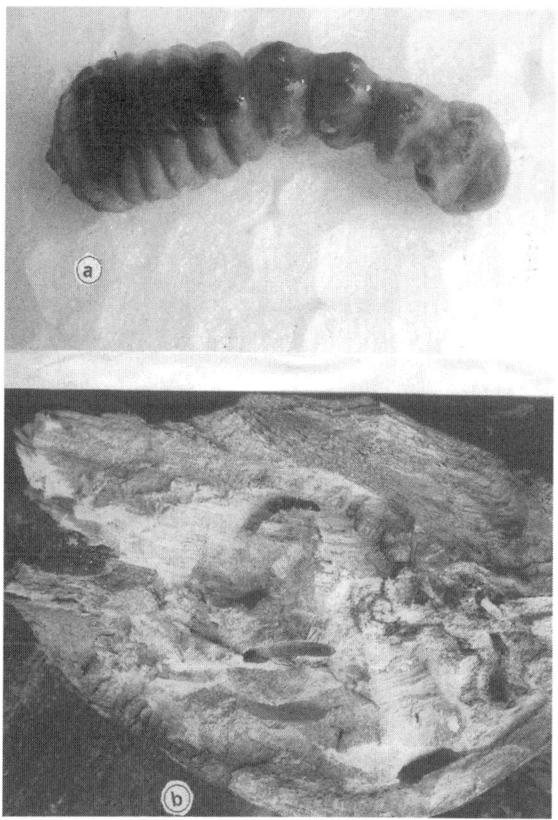

Fig. 7.6 : *Massicus unicolar* (a) Larva (b) wood damaged by larva.

Nature of Damage (Fig. 7.6 b)

Grubs bore into the wood for feeding upon wood content and females bore the stem for ovipostion.

Host Plants

Oak *Quercus griffithii*

Control Measures

i. Collection and destruction of beetles when they appear in the forest in the month of September.
ii. A cotton ball soaked in Chloroform/Petroleum or kerosine be placed in tunnel and sealed by mud for killing pest stages in side the tunnel.
iii. Dusting the crop with 10 % BHC or 5 % Aldrin/dieldrin prior to emergence of beetles from their pupal chambers and in september months.

4. The mango stem borer

Batocera titana Thoms (Fig. 7.7, 7.8)

Distribution

Tamil Nadu, Western Ghats, Maharashtra, Kerala, etc.

Marks of Identification

The beetles measures from 55 to 75 mm in body length and is yellow or greenish yellow, motled with blackish markings and or greenish yellow, mottled with blackish markings and orange spots. Its prothorax is broder than long and with sharp black spine laterally, scutellum is heart shaped and orange coloured. Antenna is on anterior edge of the head, 3rd segment is longest of all the segments. Elytra brodest basally and constricted apically, number of raised black prominences found on elytral shoulders; six or more irregular orange spots present on elytra, 3 in basal fourth and four are largest and found on disk medianly, 5 and 6 small spots placed one below. At antero lateral point of elytra there is blunt short spine.

Life Cycle

Beetles appear in July and August. Eggs are laid in wounds of the tree. After hatching the eggs, newly emerged grub tunnel

Fig. 7.7 : *Batocera titana* (Female)

Fig. 7.8 : *Batocera titana* (Male)

down into the wood and prepare long gallery parellel to the long axis of the branch/tree. Wood dust and excreta are associated with bored portion on the plant. The grub remain in the stem upto the march. The grub feed on the bast, then sapwood and the whole of the cambium layer of the tree. The full grown grub tunnel down the heart of the tree and prepare a pupal chamber which is large in size, the grub plug it and pupate in it. Pupal period is 3½ months. The beetle found fully matured in March. The pest requires one year for complition of its life cycle from egg to adult.

Nature of Damage

Grubs cause the damage to stems and trunk by boring into bast, sapwood, cambium layer and heart wood thus, affecting growth of plant and quality of wood. Greatest damage is noticed from the months, August to March.

Host Plants

Mango *Mangifera indica,* Oak *Q. grifftithii*

Control Measures

As suggested for above pest.

5. The oak meges cerambycid

Meges marmoratus Westw. (Fig. 7.9)

Distribution

Tamil Nadu, Maharashtra

Marks of Identification

The beetle measures about 70 mm in body length and 24 mm in width. It is large sized beetle, curious white or greyish white and brown marking on elytra.

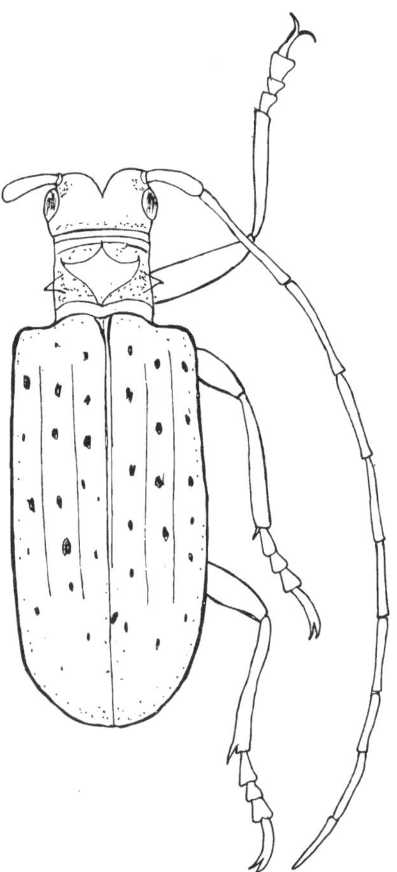

Fig. 7.9 : *Meges marmoratus*

Life Cycle

Life cycle is similar to *B. titana*

Control Measures

As above pest.

6. The bark oak weevil

Apoderus incana Stebbing (Fig. 7.10)

Fig. 7.10 : *Apoderus incana*

Distribution

North West Himalaya

Marks of Identification

The weevil measures about 7 mm in body length. It is dark yellowish brown with black marking on elytra. Rostrum is short, brown but broad and armed at end by a pair of mandibles. Its head is dark yellowish brown, short and semielliptical, narrowing posteriorly. Prothorax is with short neck, narrow anteriorly and broader posteriorly. Elytra is much broader than prothorax, legs are bright yellow. Eggs are yellow coloured and elliptical.

Life Cycle

Weevils appear in May in the forest. The female lay her eggs in the left hand corner of the apex of the leaf in May month. The leaf is then cut 2/3 across of the way down and goes to the midrib. Leaf can also be cut right across from one side, only small piece of the leaf tissue is being left on the twig. In both cases leaf get folded inwards to the midrib and then rolled up from the apex downwards. The cut portion of leaves fall down on the ground. The grub which is hatched from the egg feeds on decaying tissues and full grown within 10 days to a fortnight. The grub then leave the roll and pupate in soil.

Nature of Damage (Fig. 7.11)

Females cause damage to leaves at the time of ovipostion. Leaves cut into pieces and drop down on the ground. Thus, affecting the growth of the plant.

Fig. 7.11 : A Leaf cut by weevil

Host Plants

Ban Oak *Quercus incana* and Moru Oak *Q. dilatata*

Control Measures

i. Collection and destruction of weevils in May when they appear on the crop.

ii. Decaying leaves should be removed as part of preventive control and clean cultivation.

iii. Spraying the crop with 0.05% DDVP or 0.04% quinolophos.

7. The acorn oak weevil

Calandra sculpturata Gull. (Fig. 7.12)

Distribution

North West Himalayas

Marks of Identification

The weevil is with a curved snout and elbowed antennae. It is darked-brown in colour. Thorax is with irregular punctures. Tibiae are ribbed and with hooked spine. Elytra is about half the length of body with longitudinal punctures, it do not cover the abdomen completely. It is with broadly rounded ends. Larva is leg less, stunted, with small pale-brown head. Pupal length is equal to larval length of full grown individual. Pupa is exarate type.

Life Cycle (Fig. 7.12a, b, c, d)

Female weevil appear in June and lay her eggs immediately on the young new acrons from the preceding year's appear. The pupal stage in spring lasts for few days. From one nut as many as seven weevils have been collected. Newly emerged grub feed inside reducing the kernel to a powdery mass. No external opening is seen for emergence of matured adults. The full grown grub prepare chamber for pupation and pupate in it.

Nature of Damage

Weevils are most destructive to acrons of ban oak *Q. incana*. Grubs bore and feed upon the acrons/destroying fruits.

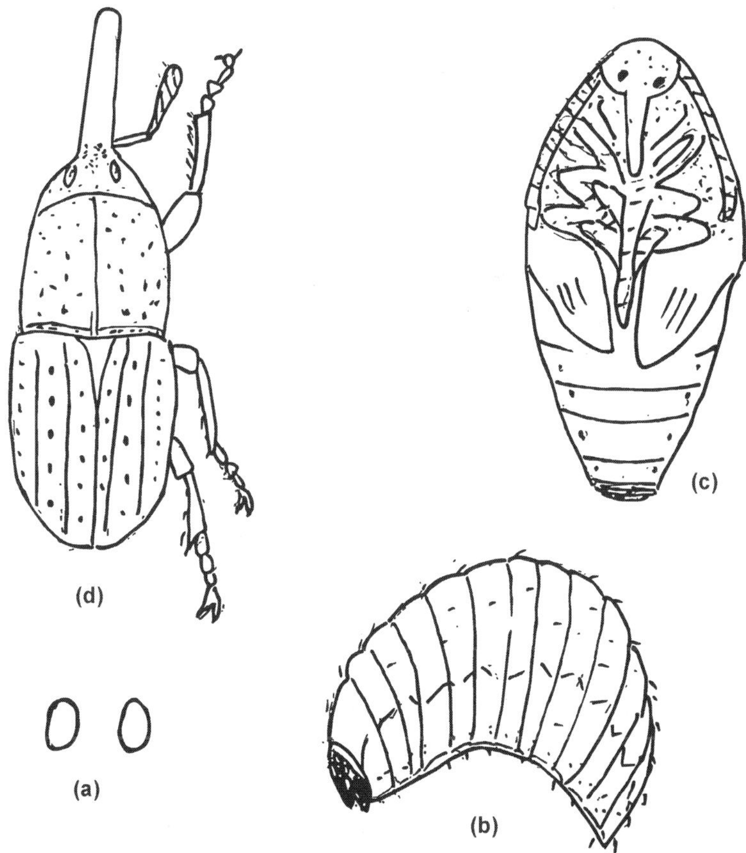

Fig. 7.12 : *Calcandra sculpturata* : Life cycle : (a) Eggs, (b) Larva, (c) Pupa (d) Adult

Host Plants

Ban oak *Q. incana*

Control Measures

i. Collection and destruction of weevils.
ii. Chemical control : As above pest.

8. The oak scolytid beetle

Chramesus globulus Stebbing

Distribution

North West Himalayas

Marks of Identification

The beetle measures about 3 mm in body length. It is globular, very convex above and flat beneath and widest across the middle. Head and thorax are black. Prothorax is pentangular in shape. Elytra of this beetle is very convex, polished or black. Abdominal segments are clothed with short spiny yellow pubescence. Under side of the beetle is flat and black.

Life Cycle

Beetles appear in first week of May at elevations of 5000 to 5500 ft. The beetle bore into wood of newly lead or dying Ban Oaks where it lay eggs. Newly emerged grubs bore straight into bark and into the sapwood. Later it bore into heart wood at an angle. The full grown grub pupate in pupal chamber and emerge as an adult in first week of May.

Nature of Damage

Beetles bore into the stem for oviposition and grubs bore the stem for feeding purpose and affect the commercial value of the timber.

Host Plants

Ban oak *Q. incana*

Control Measures

i. Removal of felled trees from forest.

ii. Removal of newly dead and dying trees from forest.

iii. Spray the felled trees with 0.05 % Nuvacron or Dusting the felled trees with 5 % Aldrin or chlordane or dusting with 10 % BHC.

9. The ban oak scolytid

Dryocoetes hewetti Stebbing.

Distribution

North West Himalayas, Kumaun, Nainital.

Marks of Identification

The beetle is very small which measures about 3 mm in body length. It has black colouration with reddish chestnut tinge. Head is convex, prothorax is broader than long. Scutellum convex and rounded. Elytra slightly broader anteriorly. Under surface of beetle is lighter reddish brown.

Life Cycle

Beetles appear in May or June. The male bore into the bast and sapwood for preparing mating chamber. It prepares sufficiently large chamber for mating. After complition of the mating chamber. Mating takes place in mating chamber which is constructed mostly right angle to the long axis of the plant. After mating, the female prepares egg gallery and she lay eggs in galleries and lastly she dies at head of the gallery. The larva hatches from the first laid egg and start tunnelling a narrow tunnel in right angle to the egg gallery. The larval galleries are straight. A full grown larva eats out a depression in the sapwood at the end of gallery and pupate in it. This is polygamous pest. The beetle feeds at night.

Nature of Damage

Males damage the tree by boring into bast and sapwood for preparing mating chambers and the grub for feeding the hard straight fibres of the oak wood. Larval galleries are not so deep in sapwood as egg galleries. In all above cases, the quality of oak wood is adversely affected.

Host Plants

Ban Oak *Q. incana*, Moru Oak *Q. dilatata.*

Control Measures

i. Small leaved trees suffer more from this pest.

ii. When the tree is full of nearly full grown larvae it should felled and barked and burnt.

iii. Trap trees should be arranged against this pest. Already attacked or sickly trees are selected and girdled or felled a week or so before flight time of the beetles.

iv. The bark should be striped off and burnt when it is infested with eggs/larvae/pupae.

v. Dusting the crop in May or June before the emergence of beetles with following pesticides 5 % Aldrin or chlordane or 10 % BHC.

10. The platypodid oak beetle

Crossotarsus fairmairei

Distribution

North West Himalayas

Marks of Identification

The dark red colured beetle measures about 5.2 mm in body length. Tarsi are of lighter brown in colour. In female the front of the head is strongly punctured. The thorax is strongly constricted at sides. In male elytral apical margin is concave and in female it is truncate.

Life Cycle

Beetles appear on wing in June month. Beetles bore into the Ban Oak plant stem for mating and oviposition purpose. The female live for some time after egg laying. Later, it dies at the opening of the tunnel and thus, blocking the exit of the tunnel. The eggs are laid at the bottom of the tunnel and grubs feed on the fungus grown on the walls of the tunnel. The larval tunnels may be 9 to 12 inch in length. The full grown individual pupate in tunnel and finally emerge as an adult from the pupa which is lasting in pupal chamber.

Nature of Damage

Beetles damage the timber by boring and by feeding upon it. The bast portion and later the heart wood is bored for feeding. The damage caused by this pest greatly reduce the market value of timber/tree.

Host Plants

Ban Oak, *Quercus incana*

Control Measures

i. The infested plant parts are to be collected and destroyed.
ii. Dusting the crop with 5% Aldrin or chlordane or 10% BHC.
iii. Spraying the crop with 0.05% DDVP or Nuacron.

11. The oak aphid

Cervaphis quercus Takahashi

Distribution

Western Ghats, India, Manipur, Thailand, etc.

Marks of Identification

Theses aphis are with many processi, marginal ones being branched and belongs to subfamily Greenidinae of family Aphididae of order Hemiptera. In apterous forms head is fused with prothorax. Antennae are 4 segmented and 5 segmented in alate forms. Antennae are shorter than body. Eyes in apterous forms are 3 faceted and in winged form they are normal. Rostrum is long, pointed at tip which extend upto the 3rd coxae. Dorsum of abdomen in non winged forms is pale brown and with deep brown transverse bands. Cornicles are long, cylindrical, curved out wards, and slightly swollen at apex. In apterous form cornide measures about 0.92 mm in length while in alate forms it is 0.82 mm. In apterous forms the aphid measures about 2.03 mm in body length and 1.10 mm in width while, in alate forms it measures about 2.02 mm in length and 0.94 mm in width.

Life Cycle

This pest reproduce parthenogenetically females and sexually males and females. In summer they produce sexuals. They have complicated life cycle and they are louse like in appearence with two pairs of transparent wings. Hind wings are half the length of fore wings, shorter than fore wings. They also shows a pair of cornicles on abdomen. This pest has complicated life cycle since it requires alternative hosts during the year to complete different generations. The life cycle has three distinct stages *viz*, egg, nymph and adult. The nymphs are very similar to adults morphologicaly but they dont show wings and they are miniatures. However, the processi are less branched in nymphal stages.

Nature of Damage

Both nymphs and adults suck the cell sap from tender leaves and shoots of the oak trees and affect the growth. The aphids secrete honey dew like substance over leaves which create sooty moulds over the leaf surface. The sooty mould affects photosynthesis of the plant and finally the growth and quality of the plant. Aphids, while sucking the cell sap inject toxins into the plant body as a result, the leaves becomes curly, they turn yellow and become dry. The flowering and fruiting bodies of the plant drop down and affect the growth vigour of the plant.

Host Plants

Oak *Quercus griffithi*

Control Measures

i. Infested plant parts should cut down and burnt along with pest stages.
ii. Aphids have natural enemies like lady bird beetles, lace wings, syrphids etc and birds like willow warbler. They cause mortalities in aphids. Encouragement should be given to above biocontrol agents.
iii. Spray the crop with following insecticides systox/ malathion/parathion/phosdrin at 0.03% or spray carbaryl 0.15% or Rogar 0.03%.

12. The oak trabala caterpillar

Trabala vishnu (Fig. 7.13)

(Lymantriidae - Lepidoptera)

Distribution

Throughout India, Western Ghats, Karnataka, Kerala, TN, Assam, West Bengal, UP, etc. Outside India, it is recorded from China, Shri Lanka, Myanmar, Java etc.

Marks of Identification

The wing expanse of male is 50-60 mm and of female is 80-90 mm. Males are pale apple green, antennae are ocherous brown. Fore wing with a faint pale antemedial line curved below the costa while on hind wing there is a pale straight oblique post medial line which becomes medial. Both wings show a series of small submarginl dark spots. Females are yellowish green. Cilia of both wings are blackish. Larval head is yellow and spotted with red. Larval colour is brownish gray and shows long lateral tufts on each segements. The first segment shows black and gray colour and others gray. Paired dorsal and lateral black spots are present on each segment. The cocoon formed by the larva is ocherous which also shows short black hairs attached to it.

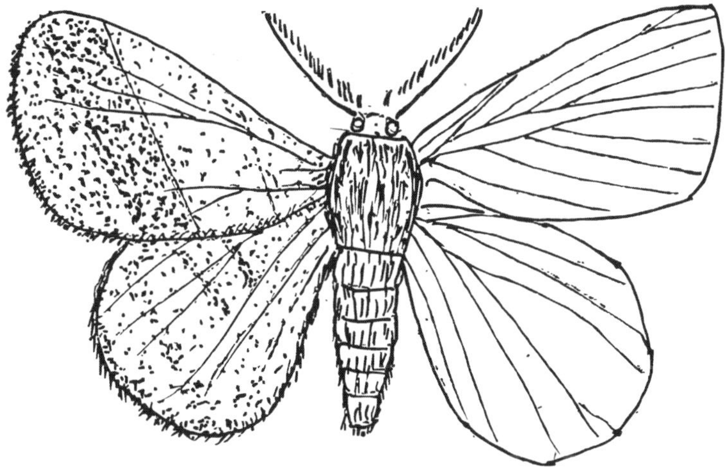

Fig. 7.13 : *Trabala vishnu* (male)

Life Cycle

Eggs are laid on leaves of the host plant. They are laid in clusters. Eggs hatch within 4-6 days. Newly emerged larvae feed on tender leaves and pass through 5 instars for full grown.

Caterpillars skeletonize the leaves consuming the cell walls and its watery contents. The full grown larva pupates in a cocoon of ocherous colour. About 35 to 45 days are requird for completion of life cycle for this pest.

Nature of Damage

Caterpillars skeletonize the leaves by feeding upon them. They consume the cell wall and watery cell contents. The loss of leaf tissue reduces the surface area of photosynthesis and food making capacity which affect the growth of plant. Weakening and stunting of growth is very common effect of the damage caused by the caterpillars.

Host Plants

Quercus incana, Q. dilatata, Q. serrata, Q. himalayan, etc.

Control Measures

i. Collection and destruction of eggs, larvae, pupae and moths.
ii. Clean cultivation and exposing pupae to natural mortality factors.
iii. Spray the crop with 0.03% DDVP or 0.03% Azadirachtin or 0.15% Carbaryl or 0.05% Endosulphan.

8

Pests of Semul (*Bombax malabaricum*)

Semul *Bombax malbaricum* is grown in Assam, West Bengal, Kerala and some other states in India. It is an important component of forestry in India. However, semul is attacked by a very large number of insect pests. The important among them are listed below.

Sr. No.	Common Name	Scientific Name	Family	Order
1.	The rubber passalid	*Leptaulax darjeelingi*	Passalidae	Coleoptera
2.	The Semul elatrid	*Ludigenus politus*	Elateridae	Coleoptera
3.	The Semul pectocera	*Pectocera cantori*	Elateridae	Coleoptera
4.	The brenthid	*Ceocephalus reticulatus*	Elateridae	Coleoptera
5.	The Semul defoliator	*Myllocerus lineatocollis*	Curculionidae	Coleoptera
6.	The scolytid	*Xyleborus* sp.	Scolytidae	Coleoptera
7.	The pyinkadu cerambycid	*Xystrocera globosa*	Cerambycidae	Coleoptera

1. The rubber passalid

Leptaulax darjeelingi

This is described under pests of India rubber.

2. The semul elaterid

Ludigenus politus

Distribution

Assam

Marks of Identification

The black coloured beetle measures about 30 mm in body length. It is elongate and narrow in shape. Its eggs and antennae are dark rufous brown. Head is with median depression. Prothorax is convex anteriorly. Elytra is black, slightly convex, sides constricted. Under surface of the beetle is dull black. The full grown larva measures about 33 mm in body length. It is elongate, dark brown above and lighter from below.

Life Cycle

Life cycle is not fully studied.

Nature of Damage

The grub cause damage to bast and sapwood by feeding upon them and making galleries. The pest is also visualized as predaceous insect of boring beetles.

Control

i. Collection and destruction of beetles.

3. The semul pectocera

Pectocera cantori Hope

This is predaceous beetle on other pests of semul.

4. The semul defoliator

Myllocerus lineatocollis

Distribution

Assam, Nilgris

Marks of Identification

The bright green or golden green coloured weevil measures for 4.5 mm to 5.5 mm in body length. Rostrum is truncate at top and moderately long. Eyes are black. Antennae are inserted at tip. Prothorax is with median longitudinal dark brown stripe. Its elytra is striate punctate. Sides straight up to hind coxa. Under surface of the weevil is golden green and finely punctate. Females are larger than males.

Life Cycle

The weevil mate in May and start egg laying in soil superficially. Eggs hatch with in 4 to 6 days. Newly emerged grubs feed on under ground parts in soil. Full grown larva pupate in soil. The adult when emerge from pupa in soil, congregate on semul tree in May-June for feeding on freshly emerged leaves.

Nature of Damage

The weevil cause damage to leaves of semul by feeding upon them and resulting in poor growth of the tree.

Host Plants

Semul *Bombax malbaricum*, Mulberry *Morus alba.*

Control Measures

i. Collection and destruction of weevils when they emerge from pupae.

ii. Digging the soil under tree for exposing immature stages of the pest to natural mortality factors.

iii. Spray the crop with any of the following insecticide.

a. DDVP - 0.05% or

b. Quinolphos 0.04%

c. Methyl parathion 0.02%

d. Malathion 0.03%

5. The pyinkadu cerambycid

Xystrocera globosa Oliv (Fig. 8.1)

Distribution

Assam, Darjeeling, TN, Maharashtra, Western Ghats, Salween river, Mussoorie, etc. Outside India it has been recorded from Myanmar, Phillipines, Egypt, Shri Lanka, Java, etc.

Marks of Identification

The beetle measures about 30 mm in body length and is reddish brown. Its elytra is testaceous yellow, with median longitunal bands diverging anteriorly, elytra is also metalic blue or green. First segment of antennae is asperate with spiniform anteior process at the apex. Its larva is bright yellowish white with brown head. Pupa is whitish, exarate type.

Life Cycle (Fig. 8.1a, b, c, d)

Beetles appear on the wing in April and lay eggs on semul or pyinkadu. After hatching the eggs, newly emerged grubs start boring into bast and then sapwood eating winding galleries too. Grub prepares broad irregular gallery for eating purpose. The full grown larva pupates in the pupal chamber prepared by itself. The beetles emerge from trees in April. Two generations are completed in a year.

Nature of Damage

The larva damages bast and sapwood by making galleries. Complete destruction to bast layer is not uncommon when larvae are numerous on the tree. In Assam semul plants are widely attacked by this pest.

Host Plants

Semul *Bombax malabaricum, Albizzia lebbek* in Egypt and pyinkadu *Xylia dolabriformis.*

Control Measures

As suggested for other cerambycid beetles.

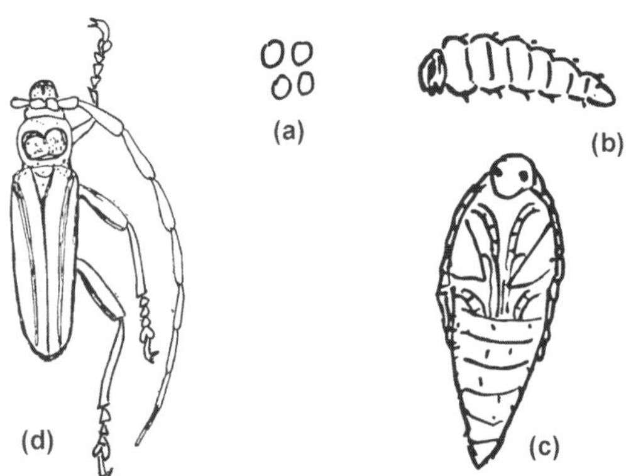

Fig. 8.1 : *Xystrocera globosa :* Life cycle : (a) eggs, (b) Larva, (c) pupa, (d) Adult (male)

9

PEST OF MAHOGANY (*SWIETENIA MAHOGANI*)

Mahogany *Swietenia mahogani* is an important forest tree in India. It is attacked by more than 2 dozon of insect pests. Some important pests are listed below.

Sr. No.	Common Name	Scientific Name	Family	Order
1.	The mahogany beelte	*Scolytoplatypus brahma*	Scolytidae	Coleoptera
2.	The mahogany scolytid	*Xyleborus gravidus*	Scolytidae	Coleoptera
3.	The mahogany buprestid	*Psiloptera fastuosa*	Buprestidae	Coleoptera
4.	The serica beetle	*Serica* sp.	Scaraeibidae	Coleoptera

1. The mahogany beetle

Scolytoplatypus brahma Blandford.

Distribution

Chittagong hill tracts

Marks of Identification

Beetles are small sized, 3 mm long. Elytra is blackish, pitchy, dull coloured. Elytra is shorter than prothorax. Antennae are brownish cetaceous, clubbed. Mid and hind legs are brown cetaceous while, fore legs are piceo - ferruginous ; under side is piceous.

Nature of Damage

The beetle cause damage to mahogany similar to *Xyleborus gravidus* by making galleries.

Host Plants

Mahogany

Control Measures

As suggested for X. *gravidus.*

2. The mahogany scolytid

Xyleborus gravidus Blandford.

Distribution

Chittagong hill tracts.

Marks of Identification

The bright reddish testaceous beetle measures about 4.5 mm in body length. Female is short robust and very convex. Head is very large, eyes are oblong, prothorax is quite broder than long, apex strongly rounded. Elytra is strongly declivous, tibiae closely and finely serrate and legs are cetaceous. Under side of the beetle is testaceous.

Nature of Damage

The insect cause the damage to mahogany tree by making galleries.

Host Plants

Mahogany *Swietenia mahogani.*

Control Measures

Treating trees with following insecticides.

i. Dusting 5% Aldrin or chlordane.

ii. Dusting 10% BHC.

3. The mahogany buprestid

Psiloptera fastuosa Fabr.

Distribution

Kolkatta, Central India, Maharashtra, Tamil Nadu, Karnataka, etc. Apart from India, it has been reported from China.

Marks of Identification

The oblong beetle measures about 18 to 26 mm in body length and about 7 mm in breadth. It has brilliant metallic coppery or deep bluish coppery colouration. Antennae are copper coloured, prothorax is 2/3 as long as wide. Elytra is medianly coppery merging into green or golden green colour. Legs and tarsi are coppery.

Life Cycle

Beetles appear on the wing in November and thus, found quite abundant. Eggs are laid on the stem. After hatching newly emerged grubs bore into stem of the tree by peeling of bark. The beetle is abundant in monsoon, specially in August and September.

Nature of Damage

Grubs bore into the stem and prepare galleries for feeding purpose. Thus, affect the growth and vitality of the plant.

Host Plants

Teak *Tectona grandis,* Mahogany *Swietenia macrophylla*.

Control

Collection and destruction of beetles, spray 0.5 % endoslphan.

4. The mahogany serica beetle

Serica sp.

Distribution

Nilambur

Marks of Identification

The beetle measures about 7.00 mm in body length. It is ovate, and reddish brownish, elytra is yellowish brown. Antennae and tarsi are brown. Prothorax is wider than long. Under surface of the beetle is brownish.

Nature of Damage

Beetles feed on leaves. They specially prefer midribs and young shoots.

Life Cycle

Not fully studied.

Host Plants

Mahogany *S. mahogani.*

Control Measures

i. Collection and destruction of beetles.

ii. Treating the crop with any one of the following insecticides

a. 0.05% Endosulphan (Spray)

b. 5% aldrin (dusting)

c. 10% BHC (dusting)

Fig. 2.3 : *Stromatium barbatum* (Adult)

Fig. 2.4 : *S. barbatum* (qrub) with bored stem.

Fig. 2.7 : *Eutectona machaeralis*

Fig. 3.1 : *Trinophyllum* sp.

(a)

(b)

Fig. 4.2 : *Helicoverpa armigera.* (a) Larva (b) Adult (Moth)

Fig. 5.1 : *Batocera rubra*

Fig. 7.5 : *Massicus unicolor* (Male)

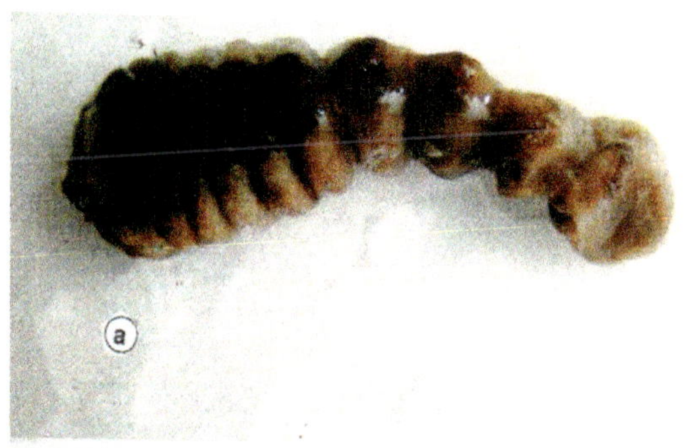

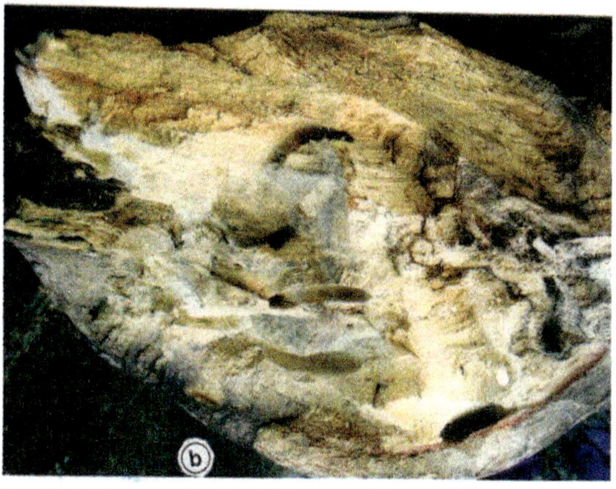

Fig. 7.6 : ***Massicus unicolar*** (a) Larva (b) wood damaged by larva.

Fig. 7.7 : ***Batocera titana*** (Female)

Fig. 7.8 : ***Batocera titana*** (Male)

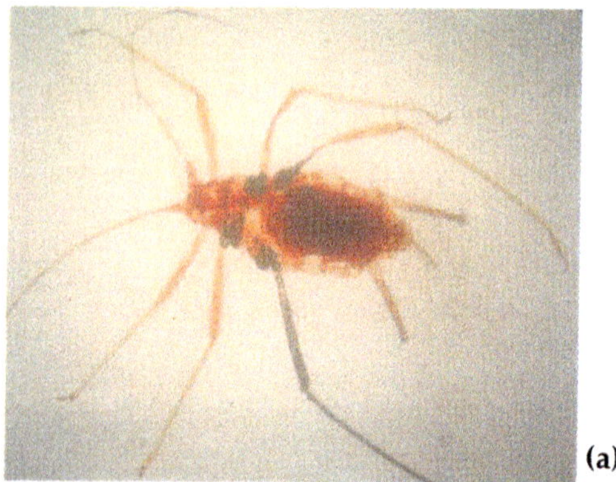

(a)

(b)

Fig. 11.1 : (a) *Aphi* sp. (b) *Chromaphis juglandicola*

Fig. 15.2 : Pine black aphid

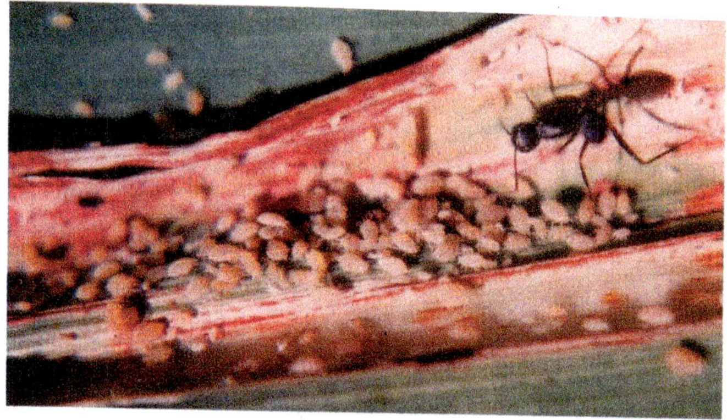

Fig. 16.1 : *Melanoaphra* sp.

Fig. 16.2 : *Ceratovacuna lanigera*

Fig. 16.3 : *Ceratovacuna bambusae*

Fig. 17.5 : *Plocaederus obesus* (Male)

Fig. 18.1 : *Coelosterna scabrata* (Female)

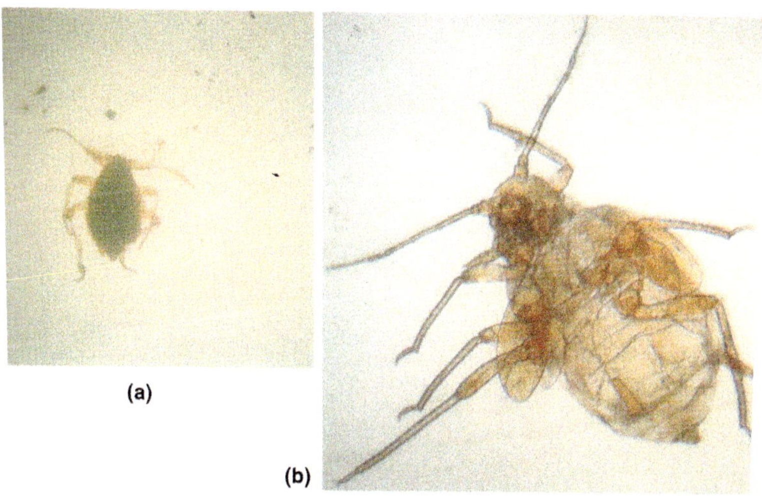

Fig. 18.2 : (a) *Aphis* sp.; (b) *Aphis craccivora*

Fig. 19.2 : (a) Twig of subabul infected with psyllids (*H. cubana*)
(b) Leaf of subabul infected with psyllids (*H. cubana*)

Fig. 20.1 : Cicada

10

Pests of Willow (*Salix* spp.)

Sr. No.	Common Name	Scientific Name	Family	Order
1.	The quetta borer	*Aeolesthes sarta*	Cerambycidae	Coleoptera
2.	The crysomelid beetle	*Melasoma populi*	Crysomelidae	Coleoptera
3.	The Lina crysomelid	*Lina populi*	Cerambycidae	Coleoptera
4.	The willow defoliator	*Lymantria obfuscata*	Lymantridae	Lepidoptera
5.	The willow cerambycid	*Xystrocera globosa*	Cerambycidae	Coleoptera

1. The quetta borer

Aeolesthes sarta Solsky (Fig. 10.1)

Distribution

Afghanistan, Turkestan, Buluchistan, Quetta, Fort sandeman, etc.

Marks of Identification

Beetles measures from 32 mm to 44 mm in body length. Antenna in male is more than twice the length of body while, in female it is less than body length. The beetle shows pubescence thickly coating elytra with glistening silvery or silvery grey colour. Prothorax is broader or angular at middle and narrower towards anterior and posterior side. The elytra is obliquely truncate and shows black, faint violet sheen or dull olive green. Both males and females produce a squeaking sound by rubbing posterior inner dorsal edge of the prothorax over anterior outer dorsal edge of mesothorax and thorax is moved up and down in verticle plane.

Eggs are milky white, elliptical and pointed at ends. Newly emerged grub measures about 1/4 inch length. Its body is yellow but head is brown and mandibles are black. Full grown grub measures about 3 inches and thick and yellowish white. Head is small but the segment connected to head is much enlarged. Spiracles are located on each side, a pair of minute legs are present on under side in each thoracic segments. Pupa is whitish yellow it shows legs, wings and antennae. The female pupa measures from 31 mm to 43 mm in body length and 9 mm to 15.5 mm in breadth.

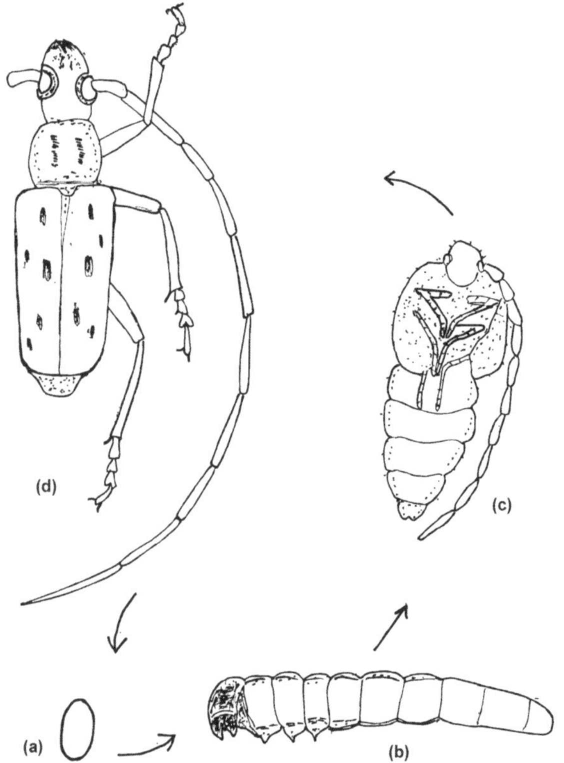

Fig. 10.1 : *Aeolesthes sarta :* Life cycle : (a) egg, (b) Larva, (c) pupa, (d) Adult (male)

Life Cycle (Fig. 10.1a, b, c & d)

Beetles appear in May/June and mate within 2-3 weeks. Mated females lay their eggs in small clumps on wounded portions left over the branches of willow. Eggs may also be laid on the bark from which the sap is oozing out or the fresh wounds on the stem, or at junction of branches or at looping point of branches. Each egg clump contain 5 to 10 eggs. About 50 eggs are laid by a single female. Egg hatching period is 5 to 10 days. Newly emerged grubs need sappy bast to feed. Therefore, the female seeks such oviposition spots so that the larva may get the immediate required food. Thus, newly emerged grub find its way into the soft sappy bark by feeding upon that content. The larva is providd with strong mandibles hence it bore deeper, upto the sapood. The sapwood is bored deeper and deeper as larva grows older. The larva prepare a winding gallery which is much broader so that the grub can turn round. The grub is voraceous feeder. Galleries have length from 9 inch to 16 inch with 4 inch breadth.

The grub works with galleries for summer and autumn and in winter, the grub bore into the heart of the tree. It is more or less horizontal work of the grub with 3 to 5 inch in length. The winter is exploited in tunnelling and eating the gallery. In spring, the grub starts with tunnelling in wood and prepare pupal chamber by the end of May. Thus, the larval period of this pest is 12 to 14 months. The pupal chamber is straight or slightly curved and comparatively narrow than other chambers. The full grown larva builds across the top, a thick white wall of the calcareous substance. The chamber is further blocked as a protection from natural enemies. The pupal period is one week or changed depending upon the climatic conditions. However, the pest avoid the danger by diapause in pupal stage for about 4 to 5 months. After emergence from the pupal chamber, beetles mate soon. The male beetle dies after mating with female. The pest take two years for completion of life cycle, 4 days egg hatching, 5.5 months in bast and sapwood, 6.5 months in wood, 5 months in pupa and same period in diapouse in beetle stage.

Nature of Damage

Both grubs and beetles cause the damage to willow plant. If natural or fresh wounds are not available on willow, the female bore into the trunk upto the sappy portion of the bark for oviposition. Likely, newly emerged grubs also feed upon sappy portion. The older grubs later bore into bast and sapwood of the tree and then into the wood for pupation. Greatest damage to the crop is done at former stages of life. In severe infestation, when beetles are planty, the beetles eat out patches and gradually join all round the stem and girdle the stem that results in the death of the tree.

Host Plants

White poplar *Populus alba,* Euphrates poplar *P. euphratica,* Kandahar willow *Salix alba,* Weeping willow *S. babylonica,* Chinar *Platanus orientalis* and Elm *Ulmus* sp. etc.

Control Measures

Preventive

i. Grubs should be deprived from the required food i.e. reaching the bast layer. Thus, early instars are crushial stages for their control.
ii. Prunning should be done in autumn and winter for avoiding thick coating of tar.
iii. Cutting and removal of infested plant parts.
iv. Adoption of trap trees for attracting beetles upon them and later, killing of them. The trap crops are immediatly removed and cut up and burnt as soon as the pest congregate on the trap crop.
v. In spring, felled trees should be removed and beetles are not allowed to oviposite upon them.
vi. Removal of infested trees in autumn.
vii. Infested tree parts are used as fire wood, this fuel must be utilized during the winter so that the emergence of future beetles would be avoided.

Chemical Control

Filling tunnels with kerosine/chloroform/petroleum soaked cotton balls and plastering with mud will kill the pest inside the tunnel.

Mechanical Control

Use iron hooks for killing and collection of pest stages from infested plant parts.

2. The willow crysomelid

Melasoma populi Linn.

Distribution

North West Himalaya

Marks of Identification

The crysomelid beetle feed on willow leaves. Grubs have been recorded in June. The larva secret strong pungent smelling fluid. The larva moult only once before pupation. Pupation takes place in June only. Beetles emerge in June 27th and survive for about one week. The larva is with black head and yellowish white body and black markings.

Nature of Damage

Beetles feed on leaves.

Host Plant

Willow *Salix* sp.

3. The lina crysomelid

Lina populi Linn.

Distribution

North West Himalaya, Shimla, Nainital, etc.

Marks of Identification

The elongate oval beetle measures about 11 mm to 12 mm in body length. It has metallic greenish blue head, prothorax, antennae and legs. Elytra is dull chrome yellow. Under surface is brillient metallic blue. Head is small. Prothorax is wider than long; Elytra is very convex.

Nature of Damage

Beetles cause damage to willow leaves by feeding upon them.

Host Plant

Willow *Salix elegans, S. babylonica.*

Control Measures

i. Collection and destruction of beetles.
ii. Dusting the crop with 5% carbaryl/aldrin or 10% BHC.

4. The willow defoliator

Lymantria obfuscata

Distribution

Shimla, Nainital, North West Himalaya, Kashmir, etc.

Marks of Identification

The male moth measures about 32 mm in wing expanse. It is grayish brown. Fore wings are with post medial double lines more regular. Hind wings with dark lunule at end of cell and dark marginal band. Males are more active than females. Females are ocherous and with a dark mark at end of cell of fore wing. The females have less aborted wings. The larvae are pale brown with short dorsal tufts of hairs and long lateral tufts. The larva also shows a dark brown dorsal band with lines down the centre and on each side. Full grown larvae measures about 40-50 mm in body length.

Life Cycle

Adults appear in June or July and the sexes mate immediately. The mated females findout suitable place on the bark of willow for oviposition. She lay her eggs in batches on the bark in June and July. Eggs are rounded, shining and light greyish brown. Each batch consists of 200 to 400 eggs and coverved by yellowish brown hairs. The winter is passed in egg stage. Eggs get hatched in March or in early April. Newly emerged larvae feed gregariously on leaves of the willow. The larva moults for 5 times and full grown within 6 to 14 weeks depending upon climatic conditions. In the afternoon, full grown larvae gathered on the bark of the tree on under surface of branches and under the debris and stones and pupate in soil among the debris. The pupal stage lasts for 9-21 days. Only one generation is completed in a year. Males survived for 4 to 10 days while, females for 11 to 31 days.

Nature of Damage

Caterpillars are the only destructive stage of the pest. They feed on willow leaves gregariously and skeletonize the plant in severe infestation. Caterpillars are voraceous feeders and they feed at night. It is serious pest of willow in Kashmir and North west Himalaya.

Host Plants

Willow *Salix elegans, S. babylonica, S. sataz* (peema), Poplar *Poplus* spp, Apricot *Prunus armeniaca* L. Apple *Pyrus malus,* Walnut *Juglans regia* L.

Control Measures

Preventive Control

i. Collection and destruction of egg masses and gregarious forms of caterpillars.

ii. Providing shelter for full grown larvae at the base of tree trunk and later, removal or collection and destruction of larvae.

iii. Clean cultivation for exposing of larvae for natural mortality factors.

iv. Digging the crop for exposing pupae for natural mortality factors.

Curative Control

Treating the crop with any of the following insecticides.

Endosulphan 0.035% or

Diamethoate 0.03% or

Phosphamidon 0.03% or

Carbaryl 0.15%

11

PESTS OF WALNUT (*JUGLANS REGIA* LINN.)

Walnut is an important temperate fruit. It is extensively cultivated in Italy, France, China and America. It is grown in Kashmir and foot hills of Himalaya, Simla, Darjeeling. In India estimated area of walnut is about 5000 hectares. Walnut fruits is rich source of protein (21%), vitamin-B and nicotinic acid. This tree is attacked by more than 30 insect pests. Important among them are listed below :

Sr. No.	Common Name	Scientific Name	Family	Order
1.	The walnut weevil	*Alcides porrectirostris*	Curculionidae	Coleoptera
2.	The long horned beetle	*Batocera horsfieldi*	Cerambycidae	Coleoptera
3.	Walnut blue beetle	*Monolepta erythrocephala*		Coleoptera
4.	Leopard moth	*Zeuzera* sp.	Cossidae	Lepidoptera
5.	Moon moth	*Actias selene*	Saturnidae	Lepidoptera
6.	Wild silkmoth	*Antheraea roylei*	Saturnidae	Lepidoptera
7.	San Jose scale	*Quadraspidiotus perniciosus*	Coccidae	Hemiptera
8.	Aphids	*Aphis pomi*	Aphididae	Hemiptera
		Chromaphis	Aphididae	Hemiptera

1. The walnut weevil

Alcides porrectirostris Marshall

Distribution

North West Himalaya, JK, Himachal Pradesh, Darjeeling.

Marks of Identification

Weevils are black in colour which measures about 8 mm to 9 mm in body length and 4 mm in breadth. The head is modified into snout. Antennae and legs are black in colour. The rostrum or snout is straight, slightly widen in females and strongly punctate in males and more finely in females. Eyes are large elliptical. Prothorax is subconical. Elytra in weevil is pubescence. Fore legs are longer than others. Femora is with fringe of long hairs and with a long tooth close to apex. The grub is whitish and legless and curved. Its head is bright yellow brown and small. The pupa is curculionid type.

Life Cycle

Mated female weevils lay their eggs on or near the female flowers of walnut in March or April or they lay their eggs on the twigs near the young flower buds in the autumn and weevils emerged in autumn only. After hatching of eggs, newly emerged grubs feed on walnut. Larvae attain the full growth in first week of July. They feed on walnut and full grown and pupate in soil. Full grown larvae drop down from the walnut and pupate in soil. The pupal stage lasts for 15 days to 20 days.

Nature of Damage

Grubs are destructive. They feed on walnut and affect the yield of the crop. From a single infected walnut normally 4 to 5 grubs are noted. In some cases 8 .to 11 grubs are noted while in others about 13 larvae are noted in single walnut. Attacked fruits start withering and the infested walnuts drop down in first week of July. Weevils hibernates in autumn under the bark or decaying leaves, stones, etc and weevil come out in March/April for proceeding their life by mating and ovipostion on the plant.

Host Plants

Walnut *Juglans regia.*

Control Measures

i. Collection and destruction of weevils by hand picking.
ii. Collection and destruction of infected walnuts.
iii. Spraying the crop with following insecticides. 0.04% quinolphos or 0.05 % DDVP or Dusting 5 to 10 % BHC early in April.

2. The long horn beetle

Batocera horsfieldi Hope

Distribution

Simla hills, Kumoon hills, Darjeeling, Kulu valley, etc.

Marks of Identification

Beetles are 45 to 65 mm long and black in colour with fine ashy or yellowish gray pubescence. Two yellow spots are present on pronotum. Elytra is with several shining black tubercles at base and rounded or broken white marks upto the apex. Full grown grub is pale yellow and 90 to 150 mm long.

Life Cycle

Mated female lay her eggs on the bark. Eggs are laid singly. Eggs hatch within 8 to 15 days. A single female can lay about 55 to 60 eggs. Newly emerged grubs bore into bark and then sapwood. The larval period is 20 to 25 months, prepupal period is 50 to 180 days and pupal period lasts for 40 to 90 days. Adult can survive for 4 to 5 months. The pest over winters in full grown larva, exists from October to March. Life cycle is completed within 23 to 32 months.

Nature of Damage

Grubs and beetles bore into the bark and wood and affect the yield of crop and quality of wood.

Host Plants

Walnut

Control Measures

i. Hand picking and mechanical destruction of grubs and beetles.

ii. Treating the holes (tunnels) with carbon bisulphide, chloroform-creosote mixture.

3. The walnut blue beetle

Monolepta erythrocephala (Baly)

Distribution

Punjab, UP, Himachal Pradesh, etc.

Marks of Identification

Beetles measure for 4 to 5 mm in body length. They are black with bluish elytra and reddish brown head. Elytra is broader at base than at prothorax. Grubs are whitish.

Life Cycle

Beetles are active from April to September. The beetle develop in soil from egg to adult.

Nature of Damage

Beetles cause damage by feeding on leaves. The damage is peaked during June to August.

Host Plants

Apple, grapevine, pear, etc.

Control measures

i. Collection and destruction of beetles.

ii. Spray the crop with 0.5% endosulphan or DDVP.

4. The leopard moth

Zeuzera sp.

Distribution

UP, HP, etc.

Marks of Identification

The moths are white with fore wing having beautiful pattern and numerous dark blue spots. Hind wings are with marginal dots. It has wing expanse of 30 to 40 mm. Caterpillars are pinkish red and 50 to 70 mm long. Eggs are oval, and dark yellow in colour.

Life Cycle

The mated female lay her eggs in cuts and wounds on the stem or in main branches singly. Eggs hatch within few days. Newly emerged caterpillars bore into tender shoots for 8 to 9 months.The thick stems or main branches are bored by older larvae. Larval period is 10 to 11 months. The full grown larva pupate inside the tunnel. Thus, life cycle is completed in about two years or more.

Nature of Damage

Caterpillars are only destructive. They bore into the bark and tender shoots and later into stems and trunk.

Host Plants

It is polyphagus pest.

Control Measures

Cotton balls soaked with chloroform/kerosine/petroleum or carbon bisulphide or paradichlorobenzene or ethyl acetate or chloroform-creosote mixture be placed in the holes and sealed with mud. The pest will die inside the tunnel.

5. San jose scale

Quadraspidiotus perniciosus (Comstock)

Distribution

Himalayan range, UP, H.P.

Marks of Identification

The insect is cell sap sucker. It is covered with black or brown covering. Its body colouration is lemon yellow. The male nymph presents elliptical or oval scale and finally develop into winged adult. Females are wingless and males are winged. Nymphs and females are flat bodied with short antennae and short 3 pairs of legs and appendages on the postsior most segment of the abdomen.

Life Cycle

A female can give birth to 200 to 400 nymphs. Eggs are laid in oviposition sac within the body of insect. Newly released nymphs are crawless. They wander here and there for settlement on tender portion of walnut tree for 12 to 24 hours. Finally they settle down on the suitable place and start sucking the cell sap. They moult for 3 to 4 times to become adult. Nymphal period is 3 to 40 days depending on climatic conditions. The male (adult) do not feed. It fertilize the females and dies. The pest overwinters in nymphal stage.

A single generation is completed within 5 to 6 weeks and several generations are completed during a year.

Nature of Damage

Both nymphs and adult females suck the cell sap from plant and while doing so inject toxins into the plant body which results in curly leaves, yellowing and drying of leaves and droping of fruiting and flowering bodies. Affect fruit setting and the yield of the crop.

Host Plants

Apple, plum, pear, peach, etc. It is polyphagus pest. It attacks about 200 species of the plants belonging to 26 families.

Control Measures

i. Collection and destruction of infested plant parts along with pest stages and burning of them.

ii. Biological control

A lady bird beetle *Rodalia cardinalis* is good biocontrol agent of san jose scale. It is mass reared and released in the field against this pest.

iii. Fumigation

Ecofriendly fumigants may be used against this pest when wind velocity is extremly low specially at evening and morning for avoiding drifting of the fumigant from target area.

6. Aphids

Chromaphis juglandicola Kalt (Fig. 11.1b)

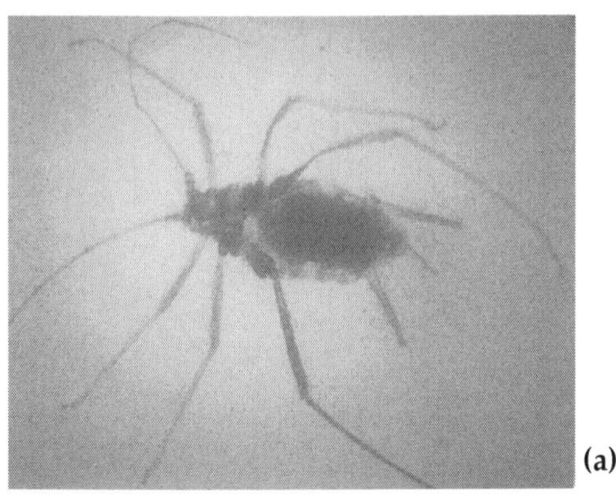

(a)

(b)

Fig. 11.1 : (a) *Aphi* sp. (b) *Chromaphis juglandicola*

Distribution

UP, H.P, Himalaya

Marks of Identification

Aphids are soft bodied insects, louse like in appearence, they are with two bars on the abdomen. Their nymphs are still smallers and don't show wings. Aphids secrete honey dew like substance and they are green coloured.

Life Cycle

The aphid reproduce parthenogenetically and viviparously. It is active from April to September. Parthenogenetically only females are produced. Hencc, they built up their population within short time. Many generations are completed in a year.

Nature of Damage

Both nymphs and adults suck the cell sap from tender parts of the plant. They confine themselves on the ventral surface of leaves and tender shoots. The affected parts become pale, leaves become curly and disfigured. The growth of plant is stunted in severe infestation. Flowering and fruiting bodies fall down on the ground. The aphid secrete honey dew like substance on leaves as result, photosynthetic activities are affected badly and growth of plant is stunted. Sooty moulds are formed on leaves. Aphids also transmit certain viral diseases to walnut.

Host Plant

Walnut, Citrus, Apple, etc.

Control Measures

i. Collection & destruction of infested plant parts along with pest stages.

ii. Encourage natural enemies like lady bird beetles, lace wings and other biocontrol agents.

iii. Use conventional insecticides such as 0.03% phosphamidon (spray) or Rogor or 0.03% or Oxydemeton methyl or quinolsphos or DDVP.

The *Aphis pomi* (Fig. 11.1a) (Aphididae : Hemiptera) (Fig. 11.1a) also cause the damage to deodara and can be controlled by 0.05% DDVP or as per above pest. Eggs are laid on shoots with large number. Eggs are hardy and adoptive against cold. Eggs hatch in spring and the pest start sucking cell sap there after on the plant.

12

PESTS OF SILVER FIR (*ABIES WEBBIANA*)

Silverfir *Abies webbiana* is grown in North West Himalayan forests. It is important forest tree of India. It is attacked by more than 25 species of the insects. Important insect pests of silver fir found in India are listed below.

Sr. No.	Common Name	Scientific Name	Family	Order
1.	The silverfir weevil	*Brachyxystus subsignatus*	Curculionidae	Coleoptera
2.	The Fir scolytid	*Xyloterus intermedius*	Scolytidae	Coleoptera
3.	The Silverfir chermid	*Chermes himalayansis*	Chermidae	Hemiptera
4.	The Silverfir Aphid	*Lachnus* sp.	Aphididae	Hemiptera
5.	The phycita caterpillar	*Phycita abietella*	Phycitidae	Lepidoptera
6.	The seed borer	*Euzophera cedrella*	Phycitidae	Lepidoptera

1. The silverfir weevil

Brachyxystus subsignatus Fst.

Distribution

Darjeeling, East Himalaya, North West Himalaya, Shimla.

Marks of Identification

The weevil measures about 4.5 mm in body length. It has chestnut brown body colour. Head is large; snout is short, bent downward with the antennae. Head is slightly smaller than thorax. Prothorax is flat with rounded sides. Elytra is strongly striate punctate. Under surface of weevil is black.

Life Cycle

Its life cycle is not fully known. Weevil may lay their eggs in soil. The grubs grow in soil with root of various plants. Pupation takes place in soil. Weevils emerge from the pupae in June. Mating takes place in June and weevils disappear from the end of June.

Nature of Damage

Weevils cause damage to new shoots of the year and thus the needles of the plant start withering and turning yellow or bright orange or die and drop off. Weevils feed on green epidermis of the new shoot of the year. A green shoot may be attacked by one or two weevils. Four or more weevils can also attack a green shoot resulting in stripping of all green content upto the woody core in the centre. In severe infestation entire loss of shoot is not uncommon. The pest attack young saplings and poles, more seriously. In general, trees of all age are attacked by this pest.

Host Plants

Spruce *Picea morinda,* Deodar *Cedrus deodara,* Silverfir *Abies webbiana.*

Control Measures

i. Collection and destruction of weevils when there is an emergance in large number in June.

ii. Spraying the crop with following insecticides.

a. quinolphos - 0.04% or

b. Methylparathion 0.02% or

c. DDVP - 0.05%

2. The silver scolytid

Xyloterus (*Trypodendon*) *intermedium* Sampson.

Distribution

North West Himalaya

Marks of Identification

Brownish beetle measures about 3.2 mm and 3.5 mm in males and females respectively. Its head is with front slightly concave, pubescent and coarsely punctate. Antenna is clubbed and club is rounded apically. Scutellum is bluntly triangular. The thorax is transverse and depressed. Elytra is more than twice as long as prothorax. In female elytra is lineate - punctate. Under side of male is more hairy than female. Legs are longer.

Life Cycle

Beetles appear in June and females girdle the branch of silver fir. Then she lay her eggs above the girdling point. Only one larva is developed from above girdle point. Girdling is made for killing the portion of branch which is above girdle. The newly hatched grubs bore into drying wood which is above the girdle. The grub bores deep and later prepare straight or curved gallery in the sapwood and going up the branch. The full grown grub enlarges the top of the gallery forming a pupal chamber and pupate in it. The gallery is blocked by wood excreta of larva. When the pupa transformed into beetle, the matured beetle findout is way outside through bark by making an exit. One generation is possible in a year but, no details are available about second generation.

Nature of Damage

i. Both beetles and grubs cause damage to fir tree. Beetles girdle the branches. The portion above girdle (ring) dies. Branches are ringed at several places.
ii. The grub bore into the branches and mines deep galleries in the sapwood.
iii. The pest attack leading shoots of seedlings, saplings and young poles.

Host Plants

Silver fir *Abies webbiana*

Control Measures

i. For nurseries and plantation, collection of twigs and branches above ring and burning them along with the pest stages is advised.

ii. Beetles are treated with 5% aldrin or 10% BHC (dusting) before they girdle the tree and when they emerge as an adluts in June.

3. The chermid

Chermes himalayansis

Distribution

Shimla, Darjeeling; East, North and West Himalaya.

Marks of Identification

Chermids are soft bodied injects measuring about less than 2 mm in body length. They have louse like appearence. They are very similar to aphids except they don't show cornicles on their abdomen. Nymphs are morphologically very simialr to adults but they lacking wings and are miniatures.

Life Cycle

Its life cycle is very complex one. It needs several alternative hosts for completion of yearly life cycle. They have specialized generations in certain seasons. They can reproduce parthenogenetically, oviparously and viviparously.

Nature of Damage

Both, nymphs and adults suck the cell sap from tender parts of silverfir and affect the growth of plant. The infested branches turn yellow and they dry. The pest also secretes honey dew like substance which create sooty moulds on the leaves and twigs and stem and affect photosynthesis and growth of the plant.

Host Plant

Spruce, Silverfir, etc.

Control Measures

i. Collection and destruction of infested twigs/branches along with pest stages.

ii. Spray the crop with 0.03% Rogor or 0.03% phosphamidon or 0.02% DDVP or 0.15 carbaryl.

13

PESTS OF POPLAR (*POPULUS CUPHRATICA*) (*POPULUS ALBA*)

Sr. No.	Common Name	Scientific Name	Family	Order
1.	The poplar buprestid	*Capnodis miliaris*	Buprestidae	Coleoptera
2.	The quetta borer	*Aeolesthes sarta*	Cerambycidae	Coleoptera
3.	The cerambycid beetle	*Prionus elliotti*	Cerambycidae	Coleoptera
4.	The willow cerambycid	*Xystrocera globosa*	Cerambycidae	Coleoptera

1. The poplar buprestid

Capnodis miliaris Klug (Fig. 13.1)

Distribution

Kashmir, Baluchistan, Syria, Persia, Tripoli, etc.

Marks of Identification

The coppery black beetle measures about 21 mm to 39.00 mm in body length. Prothorax of this beetle is whitish or gray or coppery. Head is with white squamose pubescence. Antennae are black. Prothorax is blackish, smooth with 4 black spots arranged in two rows. Elytra is black shining. Legs are black, under surface is also black. Prothorax is typically rounded or oval in shape. Full grown grub measures about 72 mm in body length. It is elongate and tapering. Its head and mandibles are black. Inverted 'V' shaped black marking present on median dorsal surface.

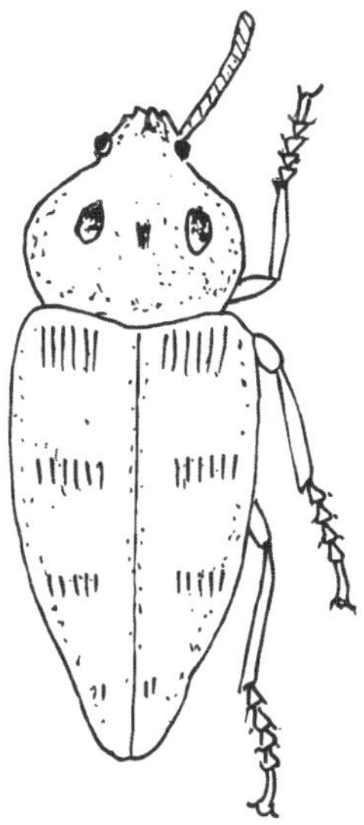

Fig. 13.1 : *Capnodis miliaris*

Life Cycle

Beetles appear in May when they emerge from diapausing stages. After mating, the female lay her eggs on poplar. On hatching, the grub bore into the stem of poplar. Grubs cut out long, shallow and broad galleries which are irregular in shape. The pupae are changed into adults in spring and summer. Thus, galleries are made in bast layer and also in sapwood and thus, causing damage to cambium layer. It is belived that there might be at least two generations in a year. However, winter is passed in the larval tunnel in the tree either in larva or pupa.

Nature of Damage

Grubs are destructive to trees. They bore bast layer, sapwood and damage cambium layer and thus affect the growth and vitality of crop.

Host Plants

Poplar *Populus euphratica, P. orientalis, P. gerardiana,* etc.

Control Measures

As like quetta borer, *A. sarta.*

2. The quetta borer

Aeolestnes sarta solsky

Details of this pest is given under pests of willow.

14

PESTS OF SPRUCE (*PICEA MORINDA*)

Spruce *Picea morinda* is grown in Indian forests specially from North West Himalaya, Deota, Garhwal, Jaunsar, Chamba state, Mussoorie, Shimla, Bashahr, Tibet Road, etc. and became characteristic of Indian forestry. However, it is attacked by a very large number of insect pests. The important insect pests recorded on spruce in Indian forests are listed below :

Sr. No.	Common Name	Scientific Name	Family	Order
1.	The spruce curculionid	*Brachyxystus subsignatus*	Curculionidae	Coleoptera
2.	The spruce weevil	*Rhyncholus himalayensis*	Curculionidae	Coleoptera
3.	The spruce scolytid	*Hylastes himalayensis*	Scolytidae	Coleoptera
4.	The blue spine scolytid	*Polygraphus major*	Scolytidae	Coleoptera
5.	The scolytid	*P. pini*	Scolytidae	Coleoptera
6.	The scolytid beetle	*Cryphalus morinda*	Scolytidae	Coleoptera
7.	The scolytid beetle	*Dryocoetes indicus*	Scolytidae	Coleoptera
8.	The Elaterid	*Lacon davidi*	Elateridae	Coleoptera
9.	The tenebrionid	*Setenis semiopaca*	Tenebrionidae	Coleoptera
10.	The tenebrionid	*Hypophloeus flavipennis*	Tenebrionidae	Coleoptera
11.	The teledapus beetle	*Teledapus dorcadioides*	Cerambycidae	Coleoptera

contd...

Sr. No.	Common Name	Scientific Name	Family	Order
12.	The cerambycid	*Leptura rubriola*	Cerambycidae	Coleoptera
13.	The cerambycid	*Tetropium* sp.	Cerambycidae	Coleoptera
14.	The scolytid	*Tomcis ribbentropi*	Scolytidae	Coleoptera
15.	The scolytid	*Pityogenes coniferae*	Scolytidae	Coleoptera
16.	The platypodid	*Crossotarsus coniferae*	Platypodidae	Coleoptera
17.	The spruce budworm	*Choristoneura fumiferana*		Lepidoptera
18.	The Gypsy moth	*Lymantria disper*	Lymantriidae	Lepidoptera
19.	The seed borer	*Phycitia abietella*	Phycitidae	Lepidoptera
20.	The seed borer	*Euzophera cedrella*	Phycitidae	Lepidoptera
21.	The chermid	*Chermes himalayensis*	Chermidae	Hemiptera
22.	The European spruce saw fly	*Diprion hercyniae*	Neodiprionidae	Hymenoptera

1. The blue pine weevil

Rhyncholus himalayensis Stebbing (Fig. 14.1)

Distribution

North West Himalaya, Shimla, Nainital, Chamba, etc.

Marks of Identification

The weevil measures about 3.8 to 4.8 mm in body length. It is black, long and narrow with short elbowed antennae. Antennal club is yellow. Head is large, very shining. Rostrum is shorter than head and stout. Prothorax is long, slightly conical and rounded on sides. Scutellum is quite small. Elytra is long, black, slightly wider than thorax, projecting down laterally below the abdomen. Its appex is constricted. Legs are short, femur is thickened in middle flat and arched. Tibiae are flattened, straight and toothed at their base. Tarsal third segment is not longer than the others and not bilobed. The body of weevil is long and parallel sided. Rostrum, head and prothorax are densely punctured.

Life Cycle

The weevil appears from May to June in considerable quantities and found boring into the wood of spruce or blue pine for egg laying. The first generation is started with the egg laying in June. Weevil's attack the trees in swarms. Weevils are found either in or beneath the bark or in sapwood. The pest bore either horizontally or at an angle into the dead bark of the standing tree. The pest later reaches sapwood boring into it. The pest works usually about for a time between the sap and bark, cutting a long groove into the bark before going into wood. The pest selects a suitable spot to bore into sapwood usually at angle and then start working for sapwood. It tunnels for an inch or less into wood and then turns to long axis of the tree for work for a couple of inches. Thus, the weevil prepares long gallery for oviposition. After oviposition the pest may not work always in straight, it may bend and again go further into the solid wood. It is belived that there might be two generations of this pest during a year.

Fig. 14.1 : ***Rhyncholus himalayensis***

Nature of Damage

The pest is capable of swarming in large numbers and by that causing severe damage to spruce trees by boring the bark and sapwood and affecting the vitality of the crop.

Host Plants

Blue pine *Pinus excelsa,* Deodar *Cedrus deodara* and spruce *Pinus morinda.*

Control Measures

i. Collection and destruction of weevils when they emerge in large quantities (swarmming).

ii. Destruction of infested parts of tree along with pest stages.

iii. Spraying the crop with any of the following insecticides.

a. DDVP - 0.05% or

b. Methylparathion 0.02% or

c. Quinolphos 0.04%

2. The spruce scolytid beetle

Hylastes himalayensis Stebbing

Distribution

North West Himalaya, Kumaun, Shimla, Chamba, etc.

Marks of Identification

Dark brown to black coloured beetle measures about 3 to 3.5 mm in body length. The beetle is elongate, shining and punctate. Head is smooth and antennae are yellow. Prothorax is constricted at appex. Elytra is wider than thorax, widest at appex; twice as long. Under surface of beetle is black with scattered whitish hairs. Legs of the beetle are dark reddish, and tibiae are red brown.

Life Cycle

Beetles appear in May and recorded upto middle of June on various trees. Second generation beetles appear in September-October. The beetle bore into bark and then into sapwood. The entrance tunnel may be horizontal to standing tree or may be an angle. The beetle tunnels the gallery and eventually the beetle reaches to the sapwood angle and this stage angle is changed. The short off set galleries are prepared by beetle and eggs are laid. The sexes mate inside the gallery made by females. The males help female in boring the position of at least of long galleries in wood. After fertilization female will continue with the heart wood.

Nature of Damage

Beetles cause damage by boring wood, bark, sapwood, etc and preparing galleries and finally ovipositing in the galleries.

Control Measures

i. Collection and destruction of insect pest stages.

ii. Encouragement of natural enemies :

a. An Ichneumonidfly which attack grubs of the pest.

b. A predatory beetle *Platysoma rimae* feed on pest.

c. *Paromalus* sp. feed on the pest.

iii. Chemical

As suggested for other scolytid beetles.

3. The spruce scolytid

Cryphalus morinda Stebbing

Distribution

North West Himalaya, Shimla, Tibet, etc.

Marks of Identification

The beetle is very small measuring about 1.70 mm in body length. It is black & thick set beetle. Its antennae and legs are yellow. Under surface is black with white hairs.

Life Cycle

Beetles appear in 2nd and 3rd week of June and they start boring with bark and cambium layer and goes to sapwood for preparing chamber. Another beetle joins to the first and both together produce narrow, small, elliptical chamber. The size of the chamber is sufficient to live for two beetles in it. This chamber is called as mating chamber. In this chamber mating process is completed and eggs are laid by the female. The female is monogamus. The grubs on hatching bore shallow gallery which is away from the mating chamber. The grub feed upon woody content of the tree. The shallow gallery later joins the larval gallery gradually under which grub can develop successfully into pupa and pupa into adult beetle.

Nature of damage

Both, beetles as well as grubs are destructive to spruce tree. The larva feed on entire cambium of the plant. The grub also bore small branches and twigs, and stem of the spruce tree. Beetles bore into the bark and reaches cambium layer. They prepare galleries in sap wood for oviposition. Two beetles works individually and finally reach to a single gallery.

Host Plants

Spruce *Picea morinda*

Control Measures

As per previous scolytid beetle.

4. The spruce dryocoetes

Dryocoetes indicus Stebbing

Distribution

North West Himalaya

Marks of Identification

The beetle is oblong, reddish to reddish brown measuring about 3.8 mm to 4.00 mm in body length. Its head is with bright yellow brush of pubescence. Antennal club is yellow. Prothorax is tapering anteriorly with small hairs. Elytra with punctures with longitudinal deep vitae. Legs are reddish brown. Under surface of beetle is dark brown.

Life Cycle

Beetles appear in September. The pest is polygamus. In a single mating chamber as much as five insects are oveserved. Beetles emerged in September start tunnelling into spruce saplings. They prepare irregular shaped depression in the cambium and sapwood in the main stem. Generally, they start working at the joints of the two branches. Beetles prepare individually a single large mating chamber of many arms usually of five. Beetles bore into the stem and reached to bast layer. In November month also beetles found working with the tree. However, beetles hibernate in the galleries from November to start of Spring. The pest is active even at the level of 9000 feet of height from sea level.

Nature of Damage

Beetles damage young green spruce saplings. Beetles riddle bast layer all round in several places and cause the death of tree.

This is serious pest of spruce. Beetles bore into stem by making large galleries reaching sapwood.

Host Plants

Spruce *Picea morinda*

Control Measures

i. Collection and destruction of beetles in September when they emerge.
ii. Chemical control : As per other scolytid beetles.

5. The spruce tenebrionid

Setenis semiopaca Blair

Distribution

North West Himalaya, Assam

Marks of Identification

The black coloured beetle measures about 37 mm in body length. Its head is slightly convex, antennae are short and capitate, prothorax is wider than long and narrower posteriorly. Its elytra is broader than prothorax, widest in apical fourth. Pygidium is short and yellow. Its legs are black, long and slender, femora is thickened. Under surface of the beetle is black and punctate.

Life Cycle

Eggs are laid on spruce tree. Grubs bore into the wood and feed upon it. The grub bore still deep for pupation in wood. Pupation takes place in pupal chamber. The adult emerge from pupal chamber and find out its way outside the spruce thereafter.

Nature of Damage

Beetles and grubs cause damage to spruce tree by boring into bark and sapwood.

Host Plants

Blue pine *Pinus excelsa*, Spruce *Picea morinda*

Control Measures

i. Collection and destruction of beetles.

ii. Collection and destruction of infested tree parts.

iii. Use of conventional pesticides

Dusting 5% Aldrin or dieldrin or

Dusting 10% BHC.

Spraying 0.15% Carbaryl

6. The spruce cerambycid beetle

Leptura rubriola Bates

Distribution

North West Himalaya, Chamba, Kashmir, etc.

Marks of Identification

Female beetle is 11-15 mm and 3.5-5 mm long and wide respectively. Its prothorax except along front and hind borders, and entire elytra is red. Antenna not extending beyond the mid of elytra. Male is black with elytra yellow red from base to little more than mid portion. Hind tarsi very long, twice the length of second and third.

Life Cycle

Female beetle lay her eggs on wounds on the tree or on the joints of two branches. Newly emerged grubs bore into the tree. The grub bore into the bark and then sapwood. When full grown it pupates inside the tunnel in stem of the spruce tree.

Nature of Damage

Grubs bore into the stem and weaken the branches and entire tree. Branches get killed one after another and finally the tree due to boring by larva and adult.

Host Plants

Spruce *Picea morinda*

Control Measures

i. Collection and destruction of beetles in June & Monsoon.

ii. Other strategies : As suggested for other cerambycids.

7. The spruce bud worm

Choristoneura fumiferana

Distribution

Western Oregon, North America, Canada

Marks of Identification

The moth measures about 3/4 inch in wing expanse. The moth varies in colouration from gray to copper. Caterpillars are pale yellowish green with black head and thoracic shield. Older instars become much darker until the general colour is brown with black markings.

Life Cycle

Eggs are greenish flattened and laid in overlapping manner in clusters. Each elongate cluster contains about 10-30 eggs or some times more. The female moth lay her eggs on needles of spruce in late summer. Egg clusters are green in colour. In cubation takes place soon after egg laying. Newly hatched larvae findout suitable places for feeding and concealment on the tree. The larva moults for five times. The full grown larva spin a light covering of silken threads around themselves and hibernates in it. Hibernating larvae emerge synchronizing the buds of fir tree and thus, they start feeding on the buds of the fir. Larvae some times prefer staminate flowers for feeding purpose. Later, larvae become leaf chewers and eat the foliage of the current year. The larval period is 3 weeks. Pupation takes place on the tree. Cocoons are not formed by this pest. However, the pupa is covered by silken threads prepared by matured larva before pupation. Adults are seen in July and early August. Mating takes place after emergence and female start laying her eggs on needles of spruce. The pest hibernates in hybernacula prepared by silken threads.

Nature of Damage

Caterpillars are only destructive. They feed on needles of spruce. Caterpillars first act as miner of old needles on spruce. While in black spruce it damage buds by feeding up on them. Staminate flowers are preferred more than needles by the larvae for feeding. Larvae further acts as leaf chewers and found eating the foliage of current year. Later, larvae web the needles together and form a crude shelter for them and thus cause damage to needles.

Host Plants

Black spruce, Donglass fir, Balsam fir, Eastern fir, etc.

Control Measures

i. Collection and destruction of egg clusters & caterpillars.
ii. Aerial application of 4 % carbaryl dusting or
iii. Spraying the crop with Azadirachtin 0.03 %

8. The gypsy moth

Lymantria disper

Distribution

It is cosmopolitan in distribution

Marks of Identification

Female moths are white with black markings and males are brown coloured and good fliers. However, female is unable to fly. Due to light weight and long hairs of first instars, they get carried for longer distance by wind current or birds or vehicles or man.

Life Cycle

Mated female lay her eggs on or near cocoons of itself. Eggs are laid in clusters and each cluster contains 400 eggs or more. Eggs are laid in July and hatches in spring. Newly emerged larvae are light in weight and provided with long hairs. Hence, they get carried for long distances. There are five instars in the larval stage. The first instar feed on foliage of oak, basswood, aspen, etc. Later instars are seen on chestnut, hemlock, pine and spruce, etc. By the end of June larvae full grown and pupate on trees or other convinient place. Pupal period is 10 days. The pest over winters in egg stage.

Nature of Damage

Caterpillars are foliage feeders. They feed on the foliage of oak, bass wood, aspen, etc. Ist instar and in later stages they feed on the foliage of pine, spruce, hemlock, chestnut, etc. and defoliate the trees completely. Caterpillars prefers broad leaved trees for feeding purpose.

Host Plants

Oak, bass wood, aspen, hemlock, spruce, pine, chestnut, etc.

Control Measures

i. Collection of egg clusters, larvae, pupae & moths and destruction of them.

ii. Spraying the crop with Azadirachtin 0.03% or carbaryl 0.15% or sevin/lead arsenate.

iii. Aerial dusting of 4% carbaryl 20 kg/ha.

9. The european spruce saw fly

Diprion hercyniae

Distribution

USA, Canada, UK, Europe

Marks of Identification

The European spruce sawfly is about 3/8 inch in body length. It is thick - waisted and provided with saw like ovipositor. It shows black with yellow markings on head, thorax and abdomen. Cocoons of this species are oval and brownish in colour. Larval body colouration is different in different instars. First, second and third instars show pale green body colouration, 4th and 5th instars are striped with 5 longitudinal white lines, while, 6th instars show no features of 5th instars, lines disappear from the larval body.

Life Cycle

Mated female lay her eggs on old spruce by making slits on needles. There are six instars in the larval stage. Larvae are characterized by having 8 pairs of prolegs to abdomen and 3 pairs of true legs to thorax. Larvae feed on old needles until recent years needle available. Full grown larvae discend down on ground for pupation and pupate in soil by constructing cocoon. In the cocoon the pest remain as prepupa in diapausing stage for several years. However, most of the individuals pupate and become adult in spring.

Nature of Damage

The pest cause damage to spruce tree in several ways.

i. The sawfly make the slits on needle for oviposition and thus needles are damaged affecting the growth and vigour of the plant.

ii. Larvae feed on needles and skeletonize the plant in severe infestation.

iii. Damage is caused by larvae by chewing and notching the needles.

Host Plants

11 species of spruce are attacked by this pest. However, white spruce is favoured most by this pest.

Control Measures

i. Collection and destruction of infested plant parts along with pest stages (Eggs, larvae, pupae)

ii. Introducing shrews in forest ecosystems because, shrews feed on pupae of this pest lasting in soil and litter.

iii. Spraying the crop with Azadirachtin 0.03%.

10. *Polygraphus longifolia* & *P. pini*

P. longifolia measures about 3 mm and *P. pini* 2.75 mm in body length. Beetles are black shining. Elytra dark chestnut brown in *P. longifolia* while in *P. pini* it is greenish pubescence. *P. pini* completes 4 generations and *P. longifolia* 3-4 generations in a year.

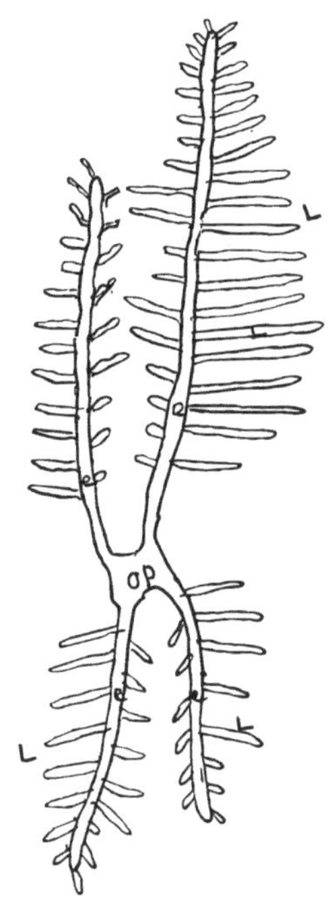

Fig. 14.2a : *Polygraphus longifolia* : damage
P - pairing chamber
L - larval gallery
e - egg gallery

Damage

Both causing damage by making galleries. Specially mating chamber or

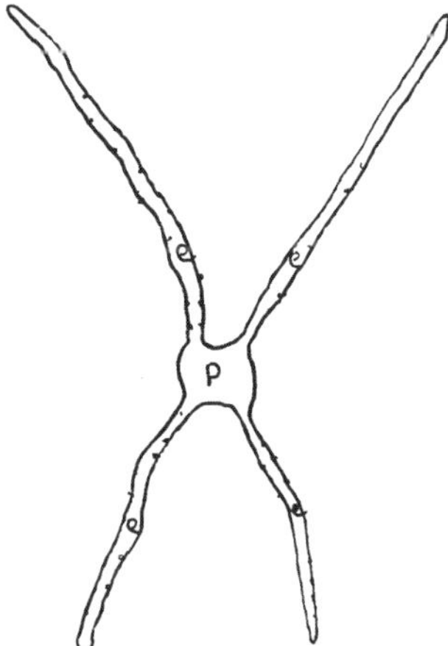

Fig. 14.2b : *Polygraphus pini :* damage : by gallerles p - pairing chamber; e - larval gallery.

pairing chamber, egg galleries and then larval galleries and finally preparing exit for adult beetles from pupal chamber. The galleries and chambers formed by the above two species are shown in figures 14.2a & b.

15

Pests of Pine (*Pinus* spp.)

In India species of pine, *Pinus longifolia,* (Chirpine), *Pinus excelsa,* Blue pine, *Pinus khasya,* etc. are grown in different forests. *Pinus longifolia* is grown in the forests of Almora, Chamba, Nainital, Kumaun, Mussoorie, Chaubattia, Bashahr, Jaunsar, Jeramola, Ravi valley and Tons valley and North West Himalaya. While other species are grown in the forest of Shillong, Simla, Chamba, N.W. Himalaya, Baluchistan, Sulieman mountains, etc. In genral, pine trees are characteristic of the Indian forests and useful source of timber and other economical products. However, these trees are attacked by a very large number of insect pests which affect the growth and quality of the crop. Followings are important insect pests which affect different species of pine in Indian forests.

Sr. No.	Common Name	Scientific Name	Family	Order
1.	The chir pine weevil	*Cryptorhynchus brandisi*	Curculionidae	Coleoptera
2.	The blue pine weevil	*C. raja*	Curculionidae	Coleoptera
3.	The pine weevil	*Rhynchous himalayensis*	Curculionidae	Coleoptera
4.	The pine scolytid	*Hylastes himalayensis*	Scolytidae	Coleoptera
5.	The pine scolytid	*H. longifolia*	Scolytidae	Coleoptera
6.	The blue pine scolytid	*Polygraphus major*	Scolytidae	Coleoptera

contd...

Sr. No.	Common Name	Scientific Name	Family	Order
7.	The blue pine scolytid	*P. pini*	Scolytidae	Coleoptera
8.	The long needle pine scolytid	*P. longifolia*	Scolytidae	Coleoptera
9.	The long needle pine smaller scolytid	*Cryphalus longifolia*	Scolytidae	Coleoptera
10.	The blue pine branch girdle	*Pityophthorus sampsoni*	Scolytidae	Coleoptera
11.	The blue pine bark borer	*Tomicus ribbentropi*	Scolytidae	Coleoptera
12.	The long needle tomicus	*T. longifolia*	Scolytidae	Coleoptera
13.	The platypodid	*Crossotarsus fairmairei*	Scolytidae	Coleoptera
14.	The chir pine borer	*Platypus biformis*	Scolytidae	Coleoptera
15.	The town ant	*Atta taxana*	Formicidae	Hymenoptera
16.	The pine wood termite	*Reticulitermes virginicus*	Termitidae	Isoptera
17.	The pine sawfly	Neodiprion spp.	Neodiprionidae	Hymenoptera
18.	The Gypsy moth	*Lymantria disper*	Lymantriidae	Lepidoptera
19.	The blue pine fruit/seed borer	*Phycita abietella*	Phycitidae	Lepidoptera
20.	The capsule borer	*Euzophera cedrella*	Phycitidae	Lepidoptera
21.	The blue pine black Aphid (Hemiptera)			
22.	The cecidomyid bud fly (Diptera)			
23.	The pyralid caterpillar (Lepidoptera)			

1. The chirpine weevil

Cryptorhynchus brandisi Steb.

Distribution

Assam, Shillong, Himalayas, Western Ghats, Myanmar, etc.

Marks of Identification

The oblong and ovate weevil measures from 6.5 to 8.5 mm in body length. It is with reddish brown elytra and black head and rostrum. Antennae are stout with swollen scape into a knob. Eyes are strongly facetted. Prothorax is triangular and tapering anteriorly. Its scutellum is round. Legs of the weevil are pubescent and punctate. The rostrum is not very thick, and tapering to apex, curved and punctate. The femora is thicken anteriorly. Under surface of the weevil is red brown. Grubs are brown and flat, yellowish white and some what curved. Pupa is yellowish, thick and blunt ended. Cocoons are blunt elliptical with rounded curved sides.

Life Cycle

Mated female lay her eggs in the bark of the pine tree by crawling deep in crevices of bark. Eggs are laid singly. Eggs hatch within few days. Newly emerged grubs remain in the thinest portion of cambium layer. The grub thus eat the tree by making a small gallery in a irregular fashion. Grubs bore deep with respect to increase of its size. It damage the thick cambium and bark by feeding upon them. The larva further bore into sapwood of pine and pupate in a pupal chamber when the beetle is fully mature, it bites out a circular hole at one end of the cocoon and crawl out of the cocoon and bores a large circular hole through the thick bark of the tree and escapes. The larval period is 6 to 8 weeks. The larva hibernates for about four months *i.e.* from December to March. The duration of life cycle vary with the region and season in Indian forest. The pest completes 3 overlapping generations in a year.

Nature of Damage

Grubs bore into the bark, cambium, bast, sapwood of pine trees. The pest attacks young green standing growth of all kinds.

Host Plants

Chir pine *Pinus longifolia and P. khasya*

Control Measures

i. The infested tree (with larvae or pupae or immature adults) should be cut down and burnt.

ii. Use trap trees in severe infestation, the trees to be felled and barked as soon as they are full of larvae.

iii. Barking of larvae.

iv. Use following natural enemies.

a. *Ichneumon* sp. (Hymenoptera), a parasitic wasp parasitizes larvae of weevil and cause mortalities in them.

b. *Cuikoo* wasp feed on grubs of the pest and cause mortalities.

c. The earwig *Elaunon bipartitus* Kirby is predaceous upon grubs of *C. brandis*.

2. The blue pine weevil

Cryptorhynchus raja Stebbing

Distribution

Himalayas, Chamba, Tehri Garhwal.

Marks of Identification

The weevil measures about 6 to 7 mm in body length. It is black, dull with very small patches and elongate narrow prothorax widest medially and tapers anteriorly. Elytra is widest at base and with small patch and spots on elytra. Legs are punctate. Rostrum tapers to middle and widens gradually, shining and brown. Antenal club is large.

Life Cycle

The life cycle of this pest is very similar to that of *C. brandis*

Nature of Damage

Grubs bore into bark, bast, cambium, sapwood and cause damage to standing trees and timber.

Host Plants

Blue pine *Pinus excelsa*, spruce

Control Measures

As per above pest

3. The blue pine scolytid beetle

Polygraphus major Stebbing

Distribution

Himalayas, Shimla, Kumaun, Garhwal, Chamba, etc.

Marks of Identification

The beetle is small, measuring about 3.2 to 3.8 mm in body length and subcylindrical and oblong. Its head and prothorax are black and shining. It has chestnut to black coloured elytra and legs. Its antenna is yellow and elytra is twice the length of prothorax. Abdomen is black. Its larva is whitish, curved and legless with yellow head. Pupa is whitish yellow. Elytral tips are typically cut down.

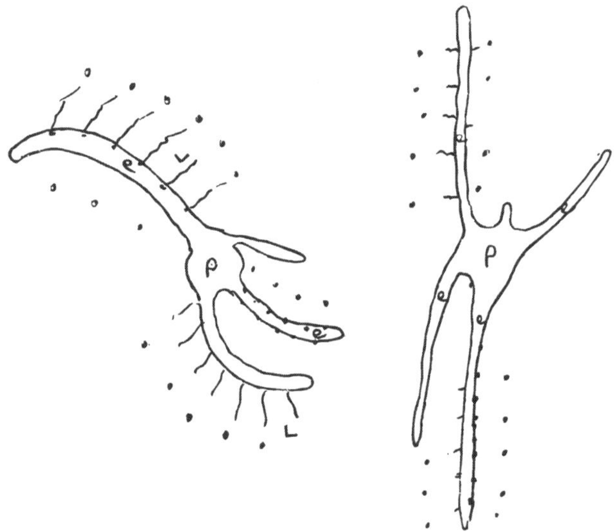

Fig. 15.1 : Plan of galleries by *P. major*
e-egg gallery, p-pairing chamber, l-larval galleries to be prepared

Life Cycle

The beetles appear on wing in May. The males bore into the bark then bast and then sapwood wherein the beetle prepares a mating chamber (pairing chamber). For oviposition, the female select only thin bark plant. This is special feature of the pest. In single mating chamber generally, three females are seen for mating purpose. However, four or five females can also be noticed in the same chamber occasionally. Thus, the male can fertilize more than 3 females in a single chamber. Each fertilized female further eat out a gallery separatelly for egg laying in separate direction. Such egg galleries are bored deep into the bast and sapwood. In egg galleries notches are prepared for filling the egg in them and 10-15 eggs are laid in single egg gallery. Oviposition days lasts for 4 to 6 weeks. Egg hatching takes place within 2 to 5 days. Newly emerged grub bore into bark and bast for feeding purpose. The grub shows very rare preference for grooving into the bast wood. Winding galleries are prepared at right angle to the egg gallery. Such tunnels measure about 1/3 to 1 inch in length which found packed with wood excreta of the grub. The larval period is about 4 weeks. Full grown larva finally eat out a small chamber into the sap wood and pupate in it. When pupa is transformed into beetle and when adult matures, it find its way out of the pupal chamber and tree by boring a tunnel as outlet. The pupal period is 8 to 13 days. The pest hibernates in winter season either in larval stage or matured beetle. The pest completes at least four generations in a single year.

Nature of Damage (Fig. 15.1)

The male beetle cause damage by boring into bark and bast for preparing mating chamber while, female eat out for preparing egg galleries. The grub feed almost entirely by boring into bast and bark and preparing winding galleries and finally pupal chamber. The pest infests young growths seriously.

Host Plants

Blue pine *Pinus excelsa,* Deodar *Cedrus deodara,* and Spruce, *Picea morinda.*

Control Measures

i. Destruction of larvae of attacked plants by cutting and destroying the infested plant parts.

ii. Attacked plant parts are removed from the forests.

iii. Use trap trees for collection and destruction of pest stages. Green trees in suitable positions should be chosen for this purpose. Small poles (green) may be felled as trap trees.

iv. Encourage the following natural enemies of the pest.

a. Use of parasitoid

The chalchid fly : It cause mortalities in grub stage.

b. Use of predaceous beetles :

Niponius canalicollis : This beetle feed on eggs and grubs of the pest and supress the popution of pests in forest.

c. The clerid beetle, *Thanasimus himalayensis* feed on the beetles of *P. major.*

d. *Hypophloeus flavipennis* (Coleoptera)

The grub of this beetle feed on the grubs of the pest.

4. The chir pine platipodid borer

Platypus biformis Chap.

Distribution

Himalayas, Darjeeling, Kumaun, Jaunsar, Chamba

Marks of Identification

The beetle is elongate, cylindrical and rufous brown which measures from 5.00 mm to 8.00 mm in body length. It has striate punctate yellowish brown elytra. Its head is as wide as prothorax and prothorax is square. In female there is a small projection just above apex on each elytron. Under surface of the beetle is shining and punctate. Larva is straight, white and elongate with abdominal segments tapering posteriorly.

Life Cycle

The mated female lay her eggs on newly felled trees and standing green trees of chirpine. The beetle prepare mating chamber or tunnel. The insect attack the trees by tunnelling straight through the bark to the sapwood and preparing zigzag gallery. Both males and females are found in the galleries. A single female can lay about 20-30 eggs. Eggs hatched within 2 days in October generation. Hatched grubs do not bore into bast or sapwood but start feeding on fungus formed in tunnels and galleries prepared by parent beetles. The life cycle from egg to adult is completed with in 6 weeks in autumn generation. The larval period is about 5 weeks. In autumn generation, grubs full grown by the end of the month of October and first week of November. The pupal period is about 2 weeks. Thus, in 3rd week of November matured beetles are released from the tree. The pest can complete 4 to 5 generations in a year.

Nature of Damage

The grub stage is not destructive in this case. The insect (beetle) causes damage to timber by boring. It cause little harm to cambium layer. Beetles cause numerous zigzag galleries into the heart wood which affect timber by weakening the log and rendering the matter useless for anything except fire wood.

Host Plants

Chir pine *Pinus longifolia,* Bamboos *Dinoderus* sp. etc.

Control Measures

i. Newly felled trees or sickly standing green trees should be removed from the forest.
ii. Badly infested plants should be cut and burnt along with pest stages.
iii. Dusting the crop with 5 % Aldrin or 10 % BHC.

5. The town ant

Atta taxana

Distribution

Western Ghats, Himalyas, Kumaun, Canada, USA, etc.

Marks of Identification

These ants are larger than black and red ants found in human dwelling. They are with petiolate abdomen and geniculate antennae. Ants are social insects. There is polymorphism, caste system and division of labour in ants. Ant colony consists 3 castes *viz,* queen, mates and workers. The town ants are soil inhabiting insects which construct their nest with a ramification of galleries and with many chambers and with hundreds of openings to the surface like a town. Hence, the name town ant. The galleries of this ant can be constructed to a depth of 15 feet or more in soil.

Life Cycle

In the first rain winged sexuals appear. They have nuptial flight for mating. Males after mating may or may not die immediately. The wings of reproductives get sheded. The fertilized female findout suitable site for nest making. The female excavate a small chamber for egg laying and isolate herself up to egg maturation and then start laying eggs. Egg hatching takes place within few days. Newly hatched grubs are whitish and legless are fed on salivary secretion of the female. When the grub is full grown it pupates in the pupal cell. Pupae does not feed. The pupae moult into adult ants. Ants can enjoy quite longer life. A queen can survive for 13 to 15 years.

Nature of Damage

Ants defoliate the trees. They cut down the leaves of pine trees and carry to their nests where leaves are chewed into pieces and used for fungus culture beds in special chambers prepared. The fungus thus grown is useful as diet for ants and is virtually pure culture. During harvesting activities the ants collect at least a single bit of leaf while they return from the trip. A single colony (large) can defoliate 4-inch tree in single day. Ants cause severe damage in fall and early spring. They also carry small buds of leaves. Needles and buds are killed due to ant damage.

Host Plants

Pine varieties

Control Measures

i. In winter, the ants congregate near the centre of nest in chamber. This is good situation for controlling the ants by fumigation. Methyl bromide is used as fumigant. The gas should be injected by tube to the depth of 2 feet in the central part of the colony.
ii. Destruction of ant nests from forest ecosystems.
iii. Killing the queens from detecting ant nests.
iv. Dusting 10% BHC on ant nests.

6. Pine saw flies

Neodiprion banksiana

Neodiprion spp.

Sawflies belong to families Tenthredinidae and Neodiprionidae of Order Hymenoptera. They cause damage to pine trees by defoliating and skeletonizing needles by feeding upon them. Several varieties of pine trees are damaged by a very large number of sawflies. Sawflies are widely attempted for their detailed studies from USA and Canada. However, no attension is paid on this important group of insects in India.

Jack pine sawfly *Neodiprion banksiana* attack older trees of pine while Red headed pine saw fly *Neodiprion lecontei* attack younger trees of pine. Many species of saw flies are strickingly very similar to each other in both appearance and habits.Larval features vary with the instars. Therefore, difficult to identify in the field condition. The characteristics of some important species are given below.

7. The jack pine saw fly

Neodiprion banksiana

Distribution

USA, Canada, India

Marks of Identification

The female sawfly measures about 3/8 inch in body length. It is yellowish brown in body colouration. The female is characterized by thread like antennae. The male is very smaller then female. It measures about 1/4 inch in body length. It is black in colour and its antennae are feather like. Larval forms are similar to caterpillars of Lepidoptera. They shows 8 pairs of abdominal prolegs and 3 pairs of walking legs (true legs). Larval forms are very dissimilar to each other in different instars hence, one can easily confused with their identification and are difficult to identify. Thus, the pest is very specialized for its sexual diamorphism and larval dissimilarties.

Life Cycle

Mated sawfly female lay her eggs on the needle of pine tree. Eggs are laid in clusters. Eggs pass the winter and hatch in May. Newly emerged larvae feed on the needles gregariously. Larvae are full grown in about 4 to 5 weeks by feeding upon needles. Full grown larvae discend down on the ground for pupation and pupate in the litter beneath the infested trees. The matured larva spin the cocoon around itself and remain in the prepupal stage until late August or September. Later, some transferred into adults while others remain in the diapausing stage until the following autumn. Jack pine sawfly infest more to the older trees than younger trees.

Nature of Damage

Grubs (Larvae) damage the pine by feeding upon needles of the tree while adult females cause by making wounds to needles at the time of oviposition. The pest attacks older pine trees than younger trees.

Host Plants

Many varieties of pine tree

Control Measures

i. Collection and destruction of egg masses and larval forms.
ii. Collection and destruction of infested plant parts along with pest stages.
iii. Clean cultivation for avoiding seltering of cocoons.
iv. Digging field underneath the tree for exposing pupae/ cocoons for natural mortality factors.
v. Use of parasitoids : Ichneumonids and Braconids.
vi. Introduction of shrews in forest ecosystem for control of this pest. Shrews digout the pupae/cocoons of the sawfly and feed upon them.

8. The red headed pine saw fly

Neodiprion lecontei

This pest attack young trees of pine. Life cycle of this pest is more or less similar to above sawfly. Nature of damage is also similar to above pest. Hence, control measures are also similar.

9. The gypsy moth

Lymantria disper

This pest is discussed under the pests of spruce

10. The pine wood termite

Reticulitermes virginicus

Distribution

India, Canada, USA, etc.

Marks of Identification

Termites are social insects. There is polymorphism, division of labour and caste system. The termite can digest cellulose. It is soft bodied, grayish white, wingless insect which resembles with ants. Therefore, termites are called as white ants. Their mandibles are very strong to feed on hard and woody content of trees or timber. The worker is the real caste which cause damage to the tree.

Life Cycle

Queen is machine of egg laying. After mating with king, the queen start egg laying. She can lay about 70,000 to 80,000 eggs in a single day. Eggs hatch within 6 to 8 days. Nymphal period is about 75-80 days. The reproductives required 1 to 2 years for full maturation. The queen can survive for 6 to 9 years and can lay continuous eggs in her life.

Nature of Damage

Workers and soldiers construct their nest on tree trunk and branches of stem. Termites feed on woody content by that damaging the bark and affecting growth and vigour and survival of the tree. Due to cellulase enzyme, the termite can digest cellulose.

Host Plants

Many varieties of pine trees and many forest plants

Control Measures

i. Termites can be controlled by sprinkling water on termiteria (Nest of termites) or the mound prepared on tree trunk. The termitaria will be collapsed by sprinkling water upon it. They do not cause damage to tree without constructing their nests on tree parts.

ii. Treating the crop with Aldrex - 30 EC or Dialdrex - 18 EC 0.1 to 0.3 % spray or Endrex - 20 EC at 0.20 % spray or chloropyriphos for treating dead and dying trees in forest.

iii. Remove the felled trees from the forest or the trees felled should be treated with above insecticides.

11. Pine black aphid (Fig. 15.2)

It suck the cell sap from the tender parts of the plant and affect the growth of tree.

Fig. 15.2 : Pine black aphid

12. The blue pine tomicus bark borer

Tomicus ribbentropi Stebbing (Fig. 15.3)

Fig. 15.3 : *Tomicus ribbentropi* (Female)

Distribution

North West Himalaya

Marks of Identification

The beetle measures about 5.5 mm in body length. It is oblong, black, shining with elytra slightly longer than thorax and striate-punctate and truncate posteriorly (Fig. 15.4). Its antennae and legs are piceous brown.

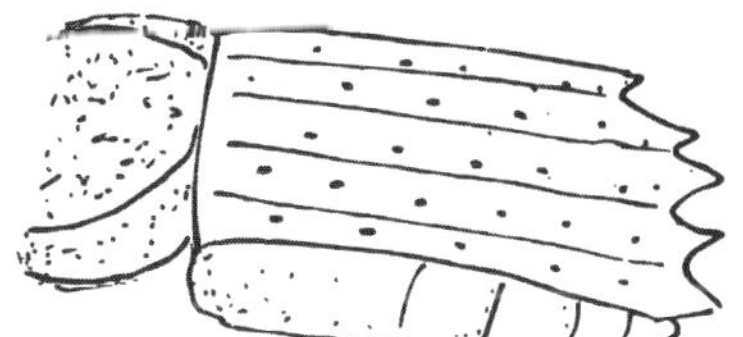

Fig. 15.4 : Eleytra and Abdomen Lateral view (T.r.)

Life Cycle

Beetles appear in April. Make pairing chamber in bast and mate in the chamber. The female prepare 2 to 4 egg galleries and lay her eggs in them. After hatching the grubs bore into the bark and sapwood. The galleries are blocked by wood dust. Full grown larva pupate in pupal chamber which is at the end of egg gallery. There are at least four generations in a year.

Nature of Damage (Fig. 15.5)

Adults cause damage by making mating chamber and the grubs by tunnelling the bark and sapwood.

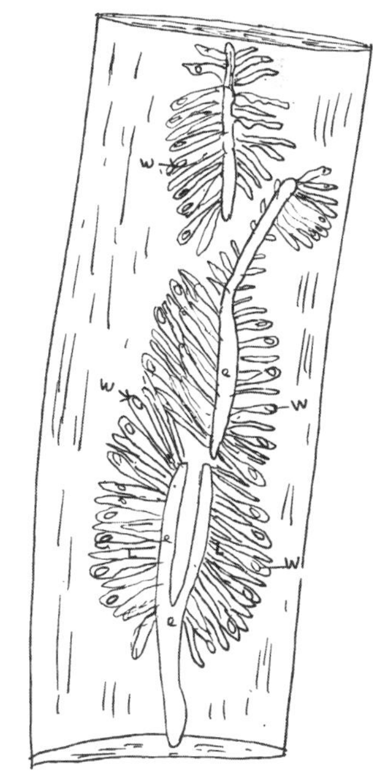

Fig. 15.5 : *Tomicus ribbentropi* : damage w-winding gallery, l-larval gallery, e-egg gallery

Host plants

Spruce *Picea morinda*, Blue pine *Pinus excelsa*

Control Measures

As per other scolotids.

16

PESTS OF BAMBOO (*DENDROCALAMUS STRICTUS*)

Bamboo is very important plant of forest and plain regions of India. Several species of bamboo are cultivated in India for commercial purpose. The important species of bamboo grown in India are *Dendrocalamus strictus. Bambusa* spp. *Bambusa polymorpha, Bambusa arundinaica. Dendrocalamus hamiltonii, Cephalostachyum pergracile,* etc. *D. strictus* is widely grown in forests of Coimbatore, Mount stuart, Dehradun, Siwaliks, Kolkatta, Raipur, Goalpara, Kachngaon, Mohnyin, Melghat forest, Annamalai Hills, Western ghats, Maharashtra, Central provinces, North India, etc. while *Bambusa* spp. are cultivated from Kolkatta region and Dehradun. *D. hamiltoni* are grown in the forests of Darjeeling while, *Cephalostachyum pergracile* is grown in Kadin bilin forest and tharrawaddy Bamboo is used for construction of houses, making furniture and goods useful in day to day life. All species of bamboos badly attacked by a large number of insect pests, which affect the growth and quality of crop. Bamboos also grown in several countries such as Brazil, Japan, Myanmar, etc. The importat insect pests of bamboo are listed in the following table :

Sr. No.	Common Name	Scientific Name	Family	Order
1.	The Bamboo Shoot Borer	*Dinoderus pilifrons*	*Bostrichidae*	*Coleoptera*
2.	The Smaller Bamboo shoot borer	*D. minutus*	*Bostrichidae*	*Coleoptera*
3.	The Bamboo bostichid	*B. brevis*	*Bostrichidae*	*Coleoptera*

contd...

Sr. No.	Common Name	Scientific Name	Family	Order
4.	The Synoxylon beetle	*Sinoxylon anale*	*Bostrichidae*	*Coleoptera*
5.	The Bostrichid beetle	*Bostrychopsis parallela*	*Bostrichidae*	*Coleoptera*
6.	The Bamboo tenebrionid	*Tribolium castaneum*	*Tenebrionidae*	*Coleoptera*
7.	The Bamboo chrysomelod	*Estigmena chinensis*	*Crysomelidae*	*Coleoptera*
8.	The Bamboo anthribid	*Anthribid* sp.	*Anthribidae*	*Coleoptera*
9.	The Bamboo weevil	*Cyrtotrachelus dux*	*Curculionidae*	*Coleoptera*
10.	Bamboo aphids	*Ceratoglyphina bambusae bengalensis*	*Aphididae*	*Hemiptera*
		Ceratovacuna bambusae	*Aphididae*	*Hemiptera*
		C.lanigera	*Aphididae*	*Hemiptera*
		Glyphinaphis bambusae	*Aphididae*	*Hemiptera*
		Melanaphis bambusae	*Aphididae*	*Hemiptera*
11.	The Bamboo Shoot Worm:	*Noctuidae*		Lepidoptera
12.	The Lymantrid Bamboo Defoiator	*Lymantriidae*		Lepidoptera
13.	The Bamboo zygaeind Moth	*Zygaeinideae*		Lepidoptera
14.	The Bamboo locust	*Aclididea*		Orthoptera
15.	The Muli Bamboo weevil	*Cyrtotrachelus longipes*-	Curculiondae-coleoptera.	

1. The bamboo shoot borer

Dinodeus pilifrons Lesne

Distribution

Western Ghats, Maharashtra, Karnataka, West Bengal, Dehradun etc.

Marks of Identification

Reddish brown beetle measures about 3.35 to 3.35 mm in body length. It has ten segmented antennae, second segment of club is rounded on inner edge. Its prothorax is rasp like anteiorly and appendages and abdominal lateral edges are lighter coloured. Basal portion of elytra is finely punctate. Light red tuffs of hair are present on front, clypeal region and inner margin of eye. Larva is yellowish white and curved with black mandibles and brownish mouth parts. Its abdominal segments are narrower then thorax, not swollen.

Life Cycle

Mated female lay her eggs in cold months specially in December or January or November. However, cold months are passed in the egg stage. The female bore into bamboo stem for mating with male and further for oviposition. Eggs are hatched in March specially in northern part of India. Newly emerged grubs feed on the woody tissue of the bamboo tree. The grub bore upward and downward in the interior. In tunnels, mass of sawdust of woody material is noticed. The full grown grub prepare a cradle by feeding on woody tissue for pupal form and pupate in it. Later, pupa is transformed into adult (beetle) which stay in the pupal chamber up to the full maturation. Then the beetle emerge out from the chamber either by preparing out let tunnel or through the out let already prepared by the parents. The pest completes over lapping 3 to 4 or more generations in a year.

Nature of Damage

Both beetles and grubs cause damage to bamboo tree by tunneling the stem. Numerous holes are seen on the nodes of bamboo. Such infections affect the quality and market value of bamboo adversely.

Host Plants

Bamboo *Dendrocalamus strictus,* other kinds of bamboos.

Control Measures

1. Collection and destruction of infected parts of bamboo tree.
2. Removal of felled trees from forests.
3. Use of predaceous beetles : *Tarsostenus univattatus* and *Tillus notatus* feed on the pest and cause mortality.

2. The smaller bamboo shoot borer

Dinoderus minutus Fabr.

Distribution

Western Ghats, Maharashtra, West Bengal, etc. Out side India, it is recoreded from Myanmar, and other tropical countries of the world.

Marks of Identification

The beetle is very small measuring about 3.00mm in body length (range 2.5 to 3.50 mm) and brownish in body colouration. Head and thorax of this beetle are black. Antennae are ten segmented with second segment of club less than one and a half times as wide as long. Prothorax shows more or less pionted teeth on anterior side. Elytra is with short reddish hairs. Larva has pale canary yellow colour and is curved.

Life Cycle

For pairing, both male and female tunnels into the bamboo and mate in mating chamber. After pairing, the female start egg laying in the interior. A single female can lay about 20 eggs. Eggs hatch within few days. Newly emerged grubs are small, white, roundish dots. Tiny grubs bore up and down in the interior of the bamboo. Larval period is about 4 weeks. The full grown grub enlarges last end of the tunnel as pupal chamber and pupate inside it and become adult and leave the tree after maturing. The pest completes more generations in a year than *D. pilifrons.*

Nature of Damage

1. Females bore into the anterior of the bamboo.
2. The hatched grubs tunnel and feed upon the wood of bamboo and undermine the strength of bamboo.
3. The beetle, after maturation, bore its way out of the bamboo tree and thus, cause the damage to plant third time by adult stage.

Host Plants

Bamboo *Dendrocalamus stictus, Bambusa* sp., *Smilax borbonica.*

Control Measures

1. Collection and destruction of infested plants of bamboo along with pest stages.
2. Bamboo should be soaked in water for longer period (5 days) and then in solution of copper sulphate and lastly soaked in water, copper sulphate and rangoon oil.
3. Oil treatment considerably prolonges the period of usefulness of bamboo
4. Use natural enemies: *Tillus notatus* - feeds on pest species.
5. Heating and smoking bamboos.
6. Bamboos should be cut on dark nights and immediately soaked or smoked for a period of two months for avoiding attack of beetles to bamboos.

3. The bamboo bostrichid beetle

Dinoderus brevis Horn

Distribution

Western Ghats, North India, West Bengal, Dehradun etc.

Marks of Identification

It differes from *D. minutus* by having more convex form and eleven segmented antennae and female is with two median teeth of the marginal row of the rasp-like anterior portion of prothorax.

Life Cycle, Damage and Control

Its life cycle, damage and control measures are similar to *D. minutus.*

4. The bamboo crysomelid

Estigmena chinensis Hope

Distribution

Coimbatore, Chennai, Western Ghats, Maharashtra.

Marks of Identification

The elongate beetle measures about 12 mm to 15 mm in body length. The head, elytra, antennae, tarsi and eyes are black. Head and thorax are narrower than elytra. Prothorax is reddish yellow or chestnut. Elytra may have different colourations as dark chestnut brown or dark reddish or black. The entire legs are brownish except the tarsi are black. However, the legs are strong and stout, Males are smaller than females, Under surface of the beetle is dark brown and smooth.

Life Cycle

Bettles appear in july. They feed upon green shoots and leaves of bamboo in july. Beetles mate in the month of July. The mated female lay ber eggs on young stem of bamboo. Only one egg is laid between two nodes. Newly emerged grubs start boring down the interior of them. Grubs also bore through nodes. Grubs are fullgrown until it reaches the lower node. Later, the grub pupates in the tunnel of stem. The pupa is developed into beetle. The full matured beetle leave the pupal tunnel by eating out exterance tunel which is partially eaten by the matured grub before pupation. In January beetles matured in internodes and leave bamboo. Beetles also appear in November in some forests. The pest completes two generations in a single year.

Nature of Damage

This is serious pest of bamboo. Beetles feed on green shoots and leaves of bamboo and affect the growth and quality of the bamboo. Grubs bore into interior of the stem thus affectinbg growth of plant and quality of bamboo and market value of the bamboo.

Host Plants

Bamboo *Dendrocalamus strictus, Dendrocalamus* sp., *Cephalostachyum pergracile.*

Control Measures

1. Collection and destruction of infested plant parts along with pest stages.
2. Dusting the crop with 5% aldrin or dieldrin or 10% BHC. This is done when beetle apperars on the crop in the months, July, January and November.

5. The dry bamboo cerambycid

Caloclytus annularis Fabr.

Distribution

Dehradun, West Bengal, Assam, North India, etc. Outside India, it has been reported from Japan, China, New Guinea etc.

Marks of Identification

The beetle measures from 10 to 15 mm in body lenth and 3 to 4 mm in breadth. It has yellow pubesence with dark brown or black markings: 3 spots on prothorax one median bifurcated posteriorly, one obliquenly oval on each side before the middle. On each elytra are two bands and a rounded spot, the first band is elliptical, the second band is transverse; submedian curved forwards along the suture about half way to base. The rounded spot lies about midway between the submedian band and the apex.

Life Cycle

Bettles appear on wing in May or June. The mated female lay her eggs in felled or cut bamboos which lost a portion of their saps (dry bamboo) in forest. In May and June beetles oviposit on felled bamboos. Eggs hatch within few days. Newly emerged grubs bore into the tissue of the walls of the bamboo and feed for tunnel from June to March. Tunnels intersect or cross the one with the other. such tunnels shows packed wood excreta and dust in them. Full grown larvae make wide tunnel as pupal chamber and pupate in it. The pupal and resting period of the beetle is about 2 months.

Nature of Damage

Grubs bore into the bamboo wood affecting growth and qulity of bamboo and market value adversely.

Host Plants

Dry Bamboos *Bambusa* spp.

Control Measures

As suggested under *D. minutus*

6. Bamboo aphids

There are about 15 species of aphids which cause damage to the bamboo plants by sucking the cell sap from the leaves and tender green shoots. The list of important species of bamboo aphids is given below.

1. *Ceratoglyphia bambusae bangalensis.*
2. *Ceratovacuna indica.*
3. *C. silvestrii.*
4. *Chaiforegma flattakana.*
5. *Astegopteryx minuta.*
6. *Pseudoasteropteryx himalayensis.*
7. *Paraoregma alexanderi.*
8. *Pseudoregma bucktoni.*
9. *Glyphinaphis bambusae.*
10. *Melanaphis bambusae*
11. *M. methalayensis*
12. *Melanaphis arundinarve*
13. *Ceratovacuna lanigera* (Fig. 16.2)
14. *Ceratovacuna bambusae* (Fig. 16.3)

7. The bamboo aphid

Ceratoglyphia bambusae bengalensis Ghosh

Distribution

Assam, Western Ghats, Kolhapur, West Bengal, Sikkim, Manipur etc.

Marks of Identification

C.b. bengalensis

Body of aphid is broadly oval. Head and thorax in apterae are fused and bear a pair of long, pointed frontal horns. Antenna is 4 or 5 segmented. Eyes in apterae 3 faceted and in alate multifaceted with ocular tubercle. Rostrum extends little beyond the fore coxae. Siphunculi poriform with hairs around base. Cauda broadly semilunar bearing many hairs. The apterous aphid measures about 2.15 mm in body length and 1.65 mm in breadth and 0.49 mm in antennal length. Its siphuncular pore is 0.05. Alate forms measure about 2.65 mm in body length and 1.46 mm in width and 0.93 mm in antennal body length, siphuncular pore is 0.05.

Life Cycle

The pest can reproduce parthenogenetically and vivipariously. The sexuals are represented by alate or apterous males and alate or apterous oviparous females. They usually form colonies at the nodal regions of young branches and many times they are also found on the lower surface of the leaves, particularly the basal region. These species are gregarious.

Host Plants

It is specific pest of bamboo only. It attack *Bambusa* sp. (graminae).

Nature of Damage

Both nymps and adults are destructive to the crop. Both suck the cell sap from young branches and leaves of the bamboo (*Bambusa* sp.) and affect the growth and vitality of plant. The pest secrete honey dew like substance on the leaves and tender shoots. Which create sooty moulds upon them. The sooty moulds affect photosynthesis of the plant and ultimmately the growth of the plant and finally the quality of bamboo. Infested leaves becomes curly, they turn yellow, become dry, flowering bodies drop down and thus affect the vigour of the plant.

Control Measures

1. Collection and destruction of infested plant parts along with pest stages.
2. Spraying the crop with malathion 0.03% or Rogor 0..3% or DDVP 0.03%
3. Use natural enemies such as lace wings, lady bird beetles, praying mantids etc.

8. The bamboo glyphinaphis

Glyphinaphis bambusae

(Hemoptera-Hormaphidinae)

Distribution

As above pest: Manipur, Meghalaya, Western Ghats, Sikkim, Arunachal Pradesh etc.

Marks of Identification

Body is oval with dark brown head, Its antennae are 4 segmented; Rostrum is short, just reaching mid coxae. Siphunculi are ring like and cauda is brown and pentagonal. Indian forms have much variations in length of ratio. It measures 1.92 mm in body length and 1.08 mm in width in alate forms and 1.72 mm in length and 1.50 mm in width in apterous forms.

Life Cycle

More or less same pattern as noted in above aphid species.

Damage

Aphids of this species form colony at nodal region and suck the cell sap. Both, numphs and adults suck the cell sap from tender parts and cause damage.

Host Plants

Bamboo and *Thysanolaena maxima* (Graminae)

9. The bamboo melanaphis

Melanaphis bambusae

Distribution

As above pest. Outside India, it has been reported from Japan, China, Malaysia, Taiwan, Australia and Igypt, etc.

Marks of Identification

Its antenna is 5 segmented in apterae and 6 segmented in alatae. Siphunculi weakly to strongly imbricated, cauda dark elongated, Femora smooth dorsally. The apterous aphid measures about 1.03 mm inbody length and 0.62 mm in width. Siphunculus is 0.11 mm long and cauda is 0.09 mm and in alate form it measures 1.33 mm in length and 1.22 mm in width. Siphunculus is 0.09 mm and cauda is 0.07mm.

Host Plants

Bambusa spp. (Graminae)

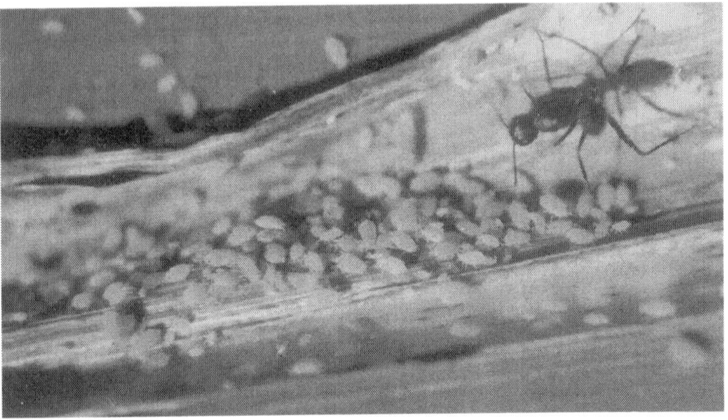

Fig. 16.1 : *Melanoaphra* sp.

Nature of Damage

Both nymphs and adults suck the cell sap from bamboo plant, particulary this species suck the cell sap from unopened leaves and lower suface of young leaves; as like other species they secrete honey dew like substance which forms the sooty moulds on leaves and affect photosynthesis and growth of the plant.

Life Cycle and Control

More or less similar to *C.b. bengalensis.*

10. Sugar cane wooly aphid

Ceratovacuna lanigera (Fig. 16.2)

Distribution

Asian countries: China, Japan, Indonesia, Taiwan, Korea, India, Pakistan, Phillipines, etc. from India it is recorded on sugar cane in Maharashtra, Karnataka, Tamilnadu, Andhra Pradesh, Orissa, UP, etc.

Marks of Identification

Adults are blackish in colour. They shows two pairs of transparent wings. Wing expanse of female is 6.50 mm. They also posses a pair of hollow tube in adults and nymphs. Both apterous and alate females are found in a colony of the aphid. They are different from other aphids by having following featues, the body is not pear shaped, the tip of abdomen is not pointed. A pair of cornicles on the abdomen are greatly reduced. It shows frontal processes or horn. The head in wingless forms is fused with thorax or with entire thorax. Newly emerged nymphs are yellowish or greenish yellow in colour. Ist instars are very active.

Fig. 16.2 : *Ceratovacuna lanigera*

Life Cycle

There are four instars in the species. Adults reproduce viviparously. Apterous (wingless) female reproduce parthenogenetically through out the year. Winged and wingless adult females are capable to give the clonal birth to nymphs (parthenogenesis). Each female produces about 35 young ones (maximum-43) within 24 hours (Hill, 1993). The ideal female produces about 215 nymphs during the period of days. Nymph takes 6 to 22 days for completion of life cycle. The life cycle is completed with in one month. Several over lapping generations are possible on sugar cane and Bamboo.

Host Plants

Sugar cane, bamboos, apple, etc.

Nature of Damage

The pest suck all the sap from tender parts of bamboo or sugarcane and affect the growth of plants. They produce sooty moulds on leaf surface which affect photosynthesis and vigour of the plant.

Control Measures

Spray the crop with carbosulphan 0.03% or 0.035% endosulphan/malathion 0.06%/DDVP 0.05%.

Biocontrol

Releasing-*Dipha aphidivora* 1000 larvae or cocoons/ha. *Micromus* sp. 2500 larvae or cocoons/ha. Syrphidfly 1000 larvae or cocoons/ha. *Chrysopa carnae*- 2500 eggs/ larvae/ha. (donot release the adults).

11. Bamboo aphid

Ceratovacuna bambusae (Fig. 16.3)

Fig. 16.3 : *Ceratovacuna bambusae*

The bamboo aphid *C. bambusae* is green coloured. It measures about 1.85 mm in length and 0.96 mm in breadth. It is greenish with while white patches. Its head is broad, dark greenish, fused to thorax. Antennae are pale green and 5 segmented and shorter than body, siphunculus with siphuncular pore. Its canda is weakly developed. Legs are dark greenish.

Life Cycle

Its life cycle is more or less similar to above species.

Host Plants

Bamboo. *Bamboo* sp.

Control Measures

Chemical : as above pest.

12. The muli bamboo weevil

Cyrtotrachelus longipes Fabr. (Fig. 16.4)

Distribution

Chittagong hill tracts.

Marks of Identification

The weevil is large sized, shining, rufousferruginous coloured with black patch on thorax and with elytral black patches. Its rostrum and legs are long. Females are much larger than males. Prothorax is slightly longer than broad. Elytra is longer than thorax. Grubs are legless, curved and whitish and pupa is whitish in colour.

Life Cycle (Fig. 16.4a, b, c, d)

Weevils appear in May/June. They mate soon and the female then findout young sporuting bamboos for egg laying. Eggs are laid on shoots, The larva on hatching bore into the tissue horizontally. It bores downward up to the base of shoot. and fullgrown. The larva pupates within the fallen buried end of the shoot at the depth of 3 to 4 inches. The pupa stay in the chamber for cold and hot seasons and emerge as an adult by the initiation of rains.

Nature of Damage

The female damages the shoots by ovipositing upon them and the grub by boring into the shoot and affecting growth and vigour of the tree.

Host Plants

Muli bamboo and other varieties.

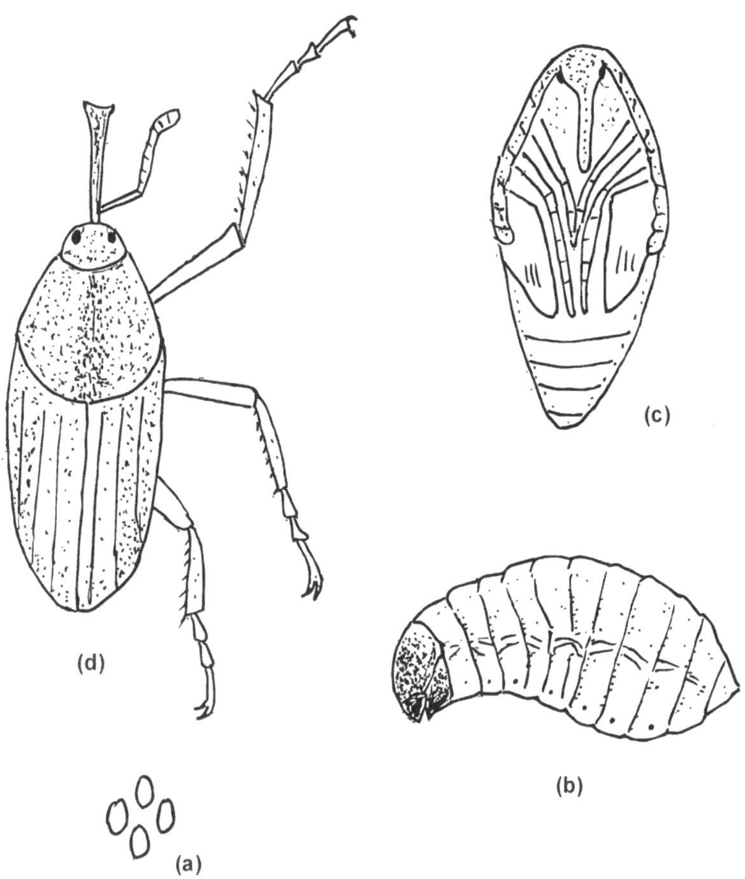

Fig. 16.4 : *Cyrtotrachelus longipes*
Life Cycle : a. Eggs; b. Larva; c. Pupa; d. Weevil

Control Measures

1. Collection and destruction of weevils & other stages.
2. Treating the crop with 5% aldrin dust or 10% BHC.

13. The bamboo shoot worm

(Noctuidae-Lepidoptera)

Distribution

Assam, Manipur, Western Ghats

Marks of Identification

Moths are stoutly builded. Caterpillars feed on tender shoots of bamboo.

Life Cycle

The mated female moth lay her eggs on bamboo shoot singly. They hatch within 4 to 6 days. Newly emerged larva feed on the tender leaves of bamboo/shoots. The larva moults for 5 times and full grown in about 14 to 17 days. The full grown larva descend on ground for pupation and pupate in soil. Pupal period is 8 to 15 days. Many generations are completed in a year.

Nature of Damage

Caterpillars are only destructive. They feed on bamboo green shoots and affect the growth of the plant.

Host Plants

Bambusia sp.

Control Measures

1. Collection and destruction of infested bamboo shoots along with caterpillars.
2. Digging the soil for exposing pupae for natural mortality factors.
3. Spray the crop with azadirachtin 0.03% or DDVP. 0.03% or endosulphan 0.05% or malathion 0.03%

2. The bamboo locust

(Acrididae-Orthoptera)

Distribution

Assam, Manipur, Sikkim, Western Ghats, West Bengal.

Marks of Identification

The locust shows jumping type of legs. Wings are straight and forewing is tegmina. It has short anternnae. It is very bad pest of forest and agricultural crops which feed on leaves and shoots by removing entire green portion.

Lify Cycle

Mated female of locust lay her eggs in soil in egg purse. Each egg purse may contain about 15-20 eggs. Eggs hatched into nymphs. Nymphs are similar to their adults but lacking wings and are smaller in size. The nymphs feed initially on bund grass and them migrate to the bamboo plant in forest for feeding on leaves of bamboo.

Nature of Damage

Both nymphs and adults defoliate the bamboo plants by feeding upon them.

Host Plants

Bambusa sp. and other varieties of bamboo.

Control Measures

1. Collection and destruction of nymphs of adults of locust.
2. Ploughing and digging soil for exposing eggs to mortality factors.
3. Blister beetle grub feed on eggs of locust in soil.
4. Spray the crop with 0.03% lindane or 0.05% chlordane or 0.1% malathion or dust 5% BHC.

17

Pests of Sal (*Shorea robusta*)

Sal *Shorea robusta* is widely grown in Indian forests. It is reported from Mandla, Siwaliks, Dholkhand, Kachugao, Gopal Para, Singbhum, Dehradun, Khandrahi, Chota Nagpur, United Provinces, Central Provinces, Balaghat, Tarai forests, Haltugaon, Kudrahi, Horai forest, Kumaun, Ganjam, Oudh sal belt, West Bengal, Assam, Anguri block, Jalpaiguri sal forest, North India, etc. Sal is very important constituent of Indian forest but attacked by more than 50 very active insect pests. Important pest insects of sal are listed below.

Sr. No.	*Common Name*	*Scientific Name*	*Family*	*Order*
1.	The passalid beetle	*Leptaulax darjeelingi*	Passalidae	Coleoptera
2.	The scarabaeid beetle	*Serica assamensis*	Scarabaeidae	Coleoptera
3.	The white qrub beetle	*Lepidiota bimaculata*	Melolonthidae	Coleoptera
4.	The white qrub beetle	*Holotricha problematica*	Melolonthidae	Coleoptera
5.	The white qrub	*H. clypealis*	Melolonthidae	Coleoptera
6.	The scarabaeid	*Heteroplia varians*	Scarabaeidae	Coleoptera
7.	The hydrophilid beetle	*Regimbartia aenea*	Scarabaeidae	Coleoptera
8.	The Nititulid beetle	*Carpophilus flavipes*	Nitidulidae	Coleoptera
9.	The Nititulid beetle	*C. hemipterus*	Nitidulidae	Coleoptera
10.	The Nititulid beetle	*Tetrisus* sp.	Nitidulidae	Coleoptera

contd...

Sr. No.	*Common Name*	*Scientific Name*	*Family*	*Order*
11.	The cucujid beetle	*Laemophloeus testaceous*	Cucujidae	Coleoptera
12.	The bostrichid	*Schistoceros anobioidea*	Bostrichidae	Coleoptera
13.	The bostrichid	*Heterobostrychus unicornis*	Bostrichidae	Coleoptera
14.	The bostrichid	*H. pileatus*	Bostrichidae	Coleoptera
15.	The bostrichid	*H. aequalis*	Bostrichidae	Coleoptera
16.	The sinosylon beetle	*Sinoxylon crassum*	Bostrichidae	Coleoptera
17.	The sinosylon beetle	*S. anale*	Bostrichidae	Coleoptera
18.	The buprestid	*Acmaeodera stictipennis*	Buprestidae	Coleoptera
19.	The buprestid	*Catoxantha bicolor*	Buprestidae	Coleoptera
20.	The buprestid	*Chrysobothris sexnotata*	Buprestidae	Coleoptera
21.	The Elaterid	*Alaus sculptus*	Elateridae	Coleoptera
22.	The tenebrionid	*Setenis laevis*	Tenebrionidae	Coleoptera
23.	The tenebrionid	*S. semivalga*	Tenebrionidae	Coleoptera
24.	The tenebrionid	*Gonocephalum depressum*	Tenebrionidae	Coleoptera
25.	The tenebrionid	*Tribolium ferrugineum*	Tenebrionidae	Coleoptera
26.	The tenebrionid	*Tribolium* sp.	Tenebrionidae	Coleoptera
27.	The tenebrionid	*Mesomorpha villiger*	Tenebrionidae	Coleoptera
28.	The chrysomelid	*Chrysomela guttata*	Crysomelidae	Coleoptera
29.	The cerambycid	*Acanthomorphus serraticornis*	Cerambycidae	Coleoptera
30.	The cerambycid	*Plocaederus obesus*	Cerambycidae	Coleoptera
31.	The cerambycid	*Hypoeschrus indicus*	Cerambycidae	Coleoptera
32.	The cerambycid	*Aeolesthes holosericea*	Cerambycidae	Coleoptera
33.	The cerambycid	*Hoplocerambyx spinicornis*	Cerambycidae	Coleoptera
34.	The cerambycid	*Xylotrechus smei*	Cerambycidae	Coleoptera
35.	The curculionid	*Conarthrus jansoni*	Curculionidae	Coleoptera
36.	The sal weevil	*Himatium asperum*	Curculionidae	Coleoptera

contd...

Sr. No.	Common Name	Scientific Name	Family	Order
37.	The scolytid	*Sphaerotrypes siwalikensis*	Scolytidae	Coleoptera
38.	The scolytid	*S. assamensis*	Scolytidae	Coleoptera
39.	The scolytid	*S. globulus*	Scolytidae	Coleoptera
40.	The scolytid	*Dryocoetes indicus*	Scolytidae	Coleoptera
41.	The scolytid	*Xyleborus fallax*	Scolytidae	Coleoptera
42.	The scolytid	*X. perforans*	Scolytidae	Coleoptera
43.	The scolytid	*X. bengalensis*	Scolytidae	Coleoptera
44.	The scolytid	*X. major*	Scolytidae	Coleoptera
45.	The platypodid	*Crossotarsus saundersi*	Platypodidae	Coleoptera
46.	The platypodid	*Platypus curtus*	Platypodidae	Coleoptera
47.	The platypodid	*Diapus furtivus*	Platypodidae	Coleoptera
48.	The platypodid	*D. quinquespinatus*	Platypodidae	Coleoptera
49.	The platypodid	*D. mirus*	Platypodidae	Coleoptera
50.	The coccid	*Monophebus* sp.	Coccidae	Hemiptera
51.	Lymantriid caterpillars	*Lymantria* spp.	Lymantriidae	Lepidoptera
52.	Inflorescence caterpillar	*Boarmia selenaria*	Lymantriidae	Lepidoptera
53.	Pyralid seed caterpillar	*Dichrocrosis leptalis*	Pyralidae	Lepidoptera
54.	The leaf caterpillar	*Ingura subapicalis*	Noctuidae	Lepidoptera
55.	Lasiocampid caterpillar		Lasiocampidae	Lepidoptera
56.	*Leucoma* sp.			Lepidoptera
57.	*Pasychira* sp.			Lepidoptera

1. The sal scarabaeid beetle

Holotricha problematica Brenske

Distribution

Kumaun, Gorakhpur, Shrinagar, etc.

Marks of Identification

Beetles measure about 16 mm in body length and 8 mm in breadth. They have dark colour. Antennae are ten segmented.

Prothorax broader than long. Scutellum is large. Larva is white, curved, wrinkled with yellow head. Full grown grub is 25 mm long.

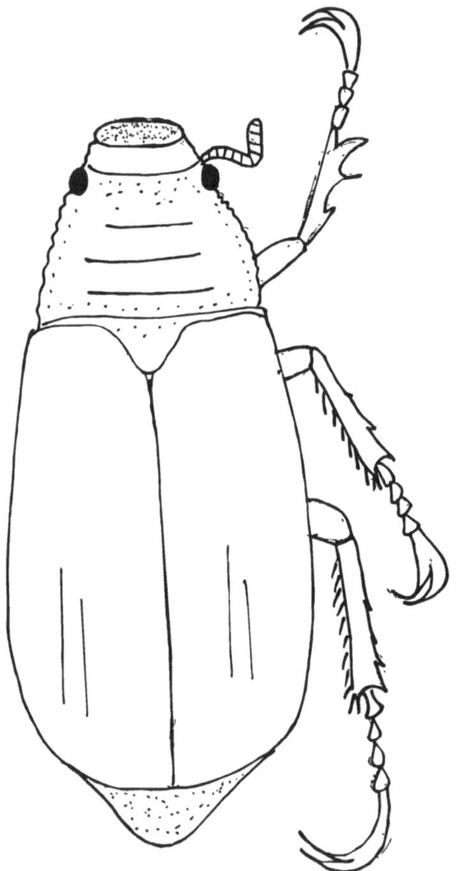

Fig. 17.1 : *Lepidiota bimaculata*

Life Cycle

Beetles appear in May on sal tree. Mated females lay her eggs in soil near sal tree in June. The larva feed in the soil on the roots of sal tree. Grubs feed on bark of roots and girdle the roots. Grubs fulfed in April and pupate in soil. Pupal stage lasts for 3 to 4 weeks. Fully developed adults emerge from soil. Beetles feed on sal tree leaves.

Nature of Damage

Grubs feed on roots of sal and beetles feed on leaves. Beetles also feed on leaves of Asan *Terminalia tomentosa,* Jaman and other species in the forest.

Host Plants

Asan *Terminalia tomentosa,* Sal *shorea robusta,* Jamun, etc.

Control Measures

i. All under growth and young trees infected are burnt.
ii. Collection and destruction of beetles (at night they emerge).
iii. Digging under ground field of the trees for exposing pest stages to natural mortality factors.
iv. Dusting the crop with 10 % BHC.
v. Use of predatory beetle : *Macrochilus bensoni* predate grubs of this pest.

L. bimaculata (Fig. 17.1)

Distribution

Maharashtra, Assam, Naga Hills, Sikkim.

Host Plants

Polyphagus

Marks of Identification

Beetle is 45-47 × 24-29 mm, elytra brown, legs black; head, prothorax, pygidium dull green.

Life Cycle

Eggs deposited in soil, grubs feed on roots of sal and soil humus and remain in soil. Pupation takes place in soil. Beetle after emergence feed at night on leaves of sal tree.

Nature of Damage

Crabs feed on roots and adults feed on leaves of sal.

Control Measures

As like *H. problematica*.

2. The sal bostrichid beetle

Schistoceros (*Caenophrada*) *anobioidea* Wt. (Fig. 17.2)

Distribution

West Bengal, Karnataka, TN, Singbhum, Chota Nagpur, Belgaum, etc. Out side India, it has been reported from Myanmar and Shri Lanka.

Marks of Identification

The beetle is 12 mm to 18 mm long and dark brown to black in colour. Prothorax is not as wide as long, anterior edge of prothorax is with bright red pubescence. Elytra is with two marginal tubercles and with very fine short rufous pubescence.

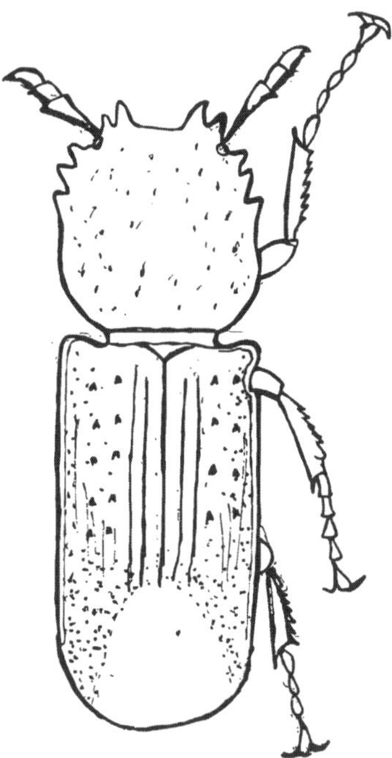

Fig. 17.2 : ***Schistoceros anobiodea***

Life Cycle

It is common pest of sal. Beetles found on the crop in March for oviposition. For egg laying beetle bores into the chamber. Newly hatched grubs start eating out ramifying galleries in the heart of the wood and develop in galleries into various instars. The full grown grub pupate inside the gallery end.

The imago develops inside the gallery. The adult beetle when mature come out of the tree by making the tunnel. Beetles can also come out from the gallery by way of an exit (hole) prepared by its campanion. The pest may have more than one generations in a year.

Nature of Damage

Both, beetles and grubs are destructive to trees. Beetles bore into the timber for oviposition and grubs bore the heart of the wood for feeding purpose and then for pupation. They can work from early March to the beginning of cold season. The pest can also damage tables, chairs, floors frequently by boring. The pest damages the above mentioned material at morning and evening.

Host Plants

Guava *Psidium guaja*, Sal *Shorea robusta* etc.

Control Measures

i. Collection and destruction of beetles.
ii. Treating the crop/Timber with following insecticides. Dusting Heptachlor/Aldrin/Chlordane at 5 kg/ha or spraying the above pesticides at 0.2 % or spraying lindane 0.2 %.

3. The sal bostrichid

Heterobostrychus aequalis

Distribution

North India and South India, specially Shimla, Dehradun, Kolkata, Chennai, Chakrata, etc. Outside India, it has been recorded from Bhutan & Myanmar.

Marks of Identification

Beetles are dark brown which measures from 6 mm to 13 mm in body length. Its head is rasp like. Its prothorax is with anterior angles often lobed. Elytra is strongly and densely punctate, Elytral teeth vary with the forms. Abdomen is with dense punctations and rasp like.

The grub is with brownish black head and whitish, curved body. Abdominal segments are corrugated. Eggs are rounded and whitish. Pupa is exarate type.

Life Cycle

Beetles develop into adult winged form in May-June. The female beetle oviposits on timber. After hatching the eggs, newly emerged grubs start eating woody content by making irregular galleries in the wood. The full grown grub pupate in tunnel and finally emerge as an adult in May/June. Only one generation is noticed during the year. Beetles could be seen in July, August and September. These beetles commonly associated with *Sinoxylon* infestation.

Nature of Damage

Grubs cause damage to timber by making irregular galleries.

Host plants

Semul *Bombax malbaricum,* Sal *Shorea robusta.*

Control Measures

i. Collection and destruction of beetles.
ii. Treating the timber with above mentioned insecticides.

4. Tha small sal buprestid

Acmaeodera stictipennis C.et.G. (Fig. 17.3)

Distribution

South India, Bangalore, Malabar, Siwalik, North Bengal, Maldah, etc.

Marks of Identification

Beetles measure about 8 mm to 10.50 mm in body length and 3 mm to 5.5 mm in breadth. Beetles are deep metallic blue or indigo blue with red elytral patch. Elytra is elongate. A red patch starts at base of outer angle and runs along the lateral margin to middle. Prothorax is finely punctate. Grubs are whitish flat and with chitinous orange prothoracic segments.

Life Cycle

Diapausing beetles emerge in June and lay their eggs on the bark of sal tree, specially on green branches or on main stems of young crop. On hatching grubs start feeding on the cambium layer and outer sap wood. Grubs grooves out winding galleries. Such galleries may have length of several inches. However, these galleries are blocked by wood dust and excreta of the grub. Full grown grubs bore deeper into the sap wood. Later, the grub prepare a small pupating chamber. Pupating chambers are clean, wood dust and larval excreta are not seen. The full grown adult find its way out of the tree by entrance tunnel found on the bark. Two generations are completed in a year. First generation is extended from June to April and second generation is completed about October/November. However, winter is passed in the grub stage of the pest.

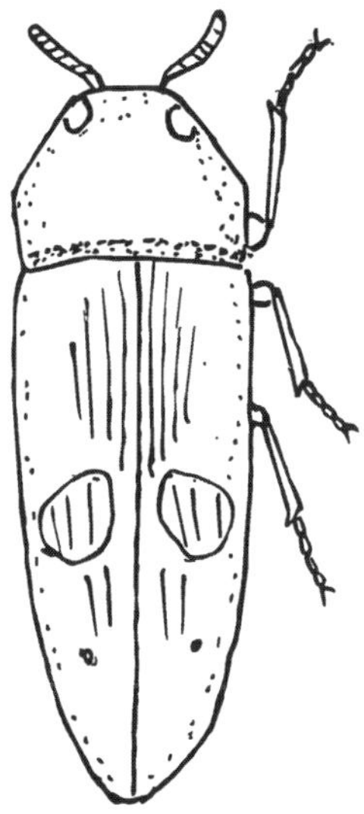

Fig. 17.3 : *Acmaeodera stictipennis*

Nature of Damage

The pest cause damage by making galleries in the stem. It attacks broad leaved trees with greatest intensity.

Host Plants

Sal *Shorea robusta*

Control Measures

i. Collection and destruction of beetles when they appear in June and later in October-November.
ii. Encouragement to parasitoids of this pest : Chalcid fly : A small fly (Hymenptera) cause mortality in grub stage by parasitism.
iii. Use of any conventional insecticide which is ecofriendly.

5. The sal buprestid

Catoxantha bicolor Fabr. (Fig. 17.4)

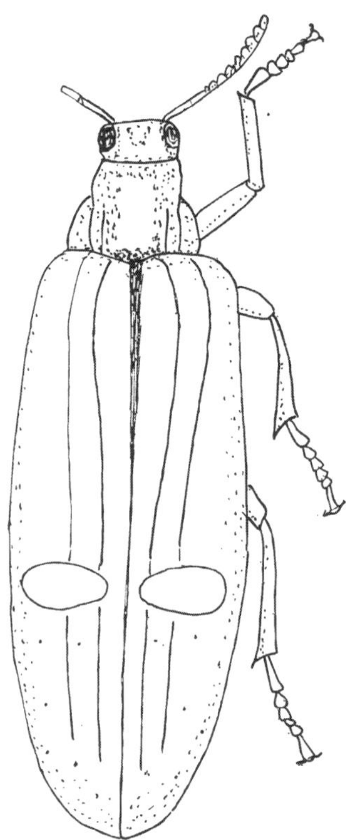

Fig. 17.4 : *Catoxantha bicolar*

Distribution

This pest is distributed in Himalaya, Darjeeling, Sikkim, Allahabad and Western Ghats. Outside India, it is reported from Myanmar, Java, Sumatra and Borneo.

Marks of Identification

Beetles are large, measuring from 55 mm to 77 mm in length and 18 mm to 25 mm in breadth. They are brilliant metallic green, blue or purple. Elytra is with two transverse oval yellow patches on apical portion. Prothorax is swollen into prominent orange yellow lobes. This species have several varieties. Grubs of this pest are elongate, yellowish white with black mandibles. Full grown grub measures about 85 mm in body length.

Life Cycle

Beetles appear on the wing in March and April in Forest of India. The female oviposits on the sal tree. On hatching the eggs, grubs bore into the bast and sapwood. The full grown grub tunnel down into the sapwood and pupate in the pupal chamber. The pupa moult into the adult in pupal chamber. The full grown beetle leave the tree (pupal chamber) and appear on the wing in March/April for new infestation.

Host Plants

Sal *Shorea robusta,* Pyinkadu *Xylia dolabriformis,* and several other forest trees.

Nature of Damage

Grubs bore into the bast and sapwood.

Control Measures

As per above pest.

6. The sal longhorn beetle

Plocaederus obesus Gahan (Fig. 17.5)

Distribution

This pest is widely distributed in Indian forests including Western Ghats.

Marks of Identification

Beetles measures from 27 mm to 45 mm in body length and 9 mm to 15 mm breadth. They are chestnut brown coloured. The antennae and legs are of same colour. The head, outer and inner edges of elytra and upper and lower edges of thorax are black in colour. First segment of antenna is large and swollen. Antennae are longer than body length. Elytral tips being truncate and spined. Tarsi are with four segments. Grubs are small and elongate but not curved. The grub shows soft yellowish head. Pupa is yellowish white with large and black eyes. Cocoons are curious calcarious whitish.

Fig. 17.5 : *Plocaederus obesus* (Male)

Life Cycle

Beetles appear on the wing in March. Mating takes place soon after the appearence of the beetle. The mated female lay her eggs on the bark in March and April. Oviposition occurs on the sickly or freshly felled trees. On hatching the eggs in April, newly emerged small grubs start feeding for a time on bast layer, making winding galleries. Grubs also eat out the sapwood. When the grub increase in size, the mandibles become stouter and hence bore down deeper into the sapwood. Grubs enjoy an internal life until they become full grown by eating sapwood and making winding galleries. Grubs remove all the bast which is below the outer bark. Later, the grub start boring deep into the sapwood. When it full grown it bore into the heart wood and make a pupal chamber for pupation. The pupal chamber is more or less at right angle to the long axis of the tree. It is observd that grubs pupate in September in curious calcarious cocoon which is the peculiar feature of this insect. In November some of cocoons get broken due to formation of beetles in them. They are matured, but unable to fly due to soft covering. The maturing individuals rests in cocoon between December and March. The pest completes only one generation in a single year.

Nature of Damage

The grub cause the damage to bast layer and timber of the tree. The larva prepare winding galleries by boring. They also prepare enlarged pupal chambers by boring. The damage to timber is more important than standing trees in forest. The pest can cause death of the tree by boring. In certain places, the beetles can kill green standing trees if infestation is severe.

Host Plants

Mango *Mangifera indica*, Sal *Shorea robusta*, Semul.

Control Measures

i. Collection and destruction of beetles, pupae and grubs.

ii. Removal of dying and sickly trees from the forests.

iii. Bark all fellings (poles, mature timber)

iv. If it is not possible to bark the felled trees, they should be removed from forest.

v. Encourage predaceous grubs of colydiid which predates larvae and pupae of the pest.

7. The sal cerambycid beetle

Hoplocerambyx spinicornis Newman

Distribution

Western Ghats, Kolhapur, Satara, Pune, Assam, West Bengal, Chota Nagpur, Allahabad, etc. Outside India this pest has been reported from Afganistan, Nepal, Sumatra, Singapore, Borneo, Phillipines, etc.

Marks of Identification

Beetles measures from 20 mm to 60 mm in body length and 5 mm to 16 mm in breadth. They are elongate black or brownish black with prominent head and long antennae. Elytra is different coloured from piceous to reddish brown. Immature forms found in pupal chamber are reddish in colour. Head, prothorax, antennae, legs and under surface are covered with greyish pubescence. In male antennae are longer than body and in female antennae are shorter than body. The female shows a long, flat, blunty, pointed, tapering, telescopic ovipositor which is about half an inch. Eggs are elongate, cylindrical and slightly swollen at anterior end. Larvae are yellowish and thickest anteriorly.

Life Cycle

Beetles appear at the end of March or begining of April in forest and then becomes scare or disappear. Later, again in september October the beetles found on the wing. Beetles fly at night and during day time they hide beneath the bark or in under growth of trees. Beetles oviposit on newly felled trees or standing trees in the forest. Mating takes place immediately in the environment irrespective of size (small or large) of males.

With the help of ovipositor, female lay her eggs in crevices of bark or on the bark. Ovipositor is used by the female for examining suitable place for ovipostion. Eggs are laid singly. A single mated female can lay several hundred eggs. Eggs hatch with in 5 to 8 days after oviposition. Newly hatched grub measures about eighth of an inch in length and full grown grub about quarter of an inch in body length. Grubs bore into bast and then into sapwood for feeding on the same matter. They prepare galleries. The gallery is at first narrow and with two or more arms. Galleries are entirely packed with wood excreta of grub. Infact, the grub is packed in the galleries with wood excreta ejected by the grub. Grubs further bore into heart wood for preparing pupal chamber. The pupal chamber is covered with white calcareous cocoon/pupal chamber. The pupal stage lasts for 2 to 3 months. The pupal period is fluctuated with respect to time, season and climate. The pest hibernates in pupal stage in cold weather.

Nature of Damage

Grubs bore into bast, sapwood and heart wood and thus, affecting the quality of timber and growth of plant in case of standing plants. Grubs can bore roots, main stem and larger branches of the tree. The cambium layer of the plant is usually completely destroyed.

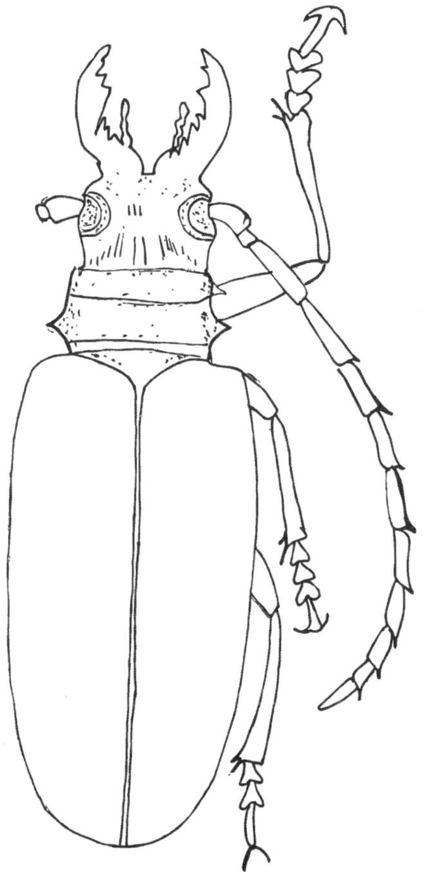

Fig. 17.6 : *Acanthophorus serraticornis* (Male)

Host Plants

Sal *Shorea robusta, Duabanga sonnatiodes* and *Pentacme suavis.*

Control Measures

i. Removal of felled trees from forest.

ii. Barking every tree within a week of felling.

iii. Collection and destruction of beetles when they appear on wing and for infestation on the tree.

iv. Encouragement of natural enemies like birds, parasitoids and predators :

a. Wood peckers feed on grubs by tunnelling the wood.

b. *Ichneumon* sp. (Hymenoptera : Ichneumonidae) cause mortalities in grub stage of the pest.

c. *Bothrideres* sp. (Coleoptera). Grubs of this predatory beetle feed on the grubs of the pest.

v. Treating felled trees with 5% Aldrin/Chlordane dusting or dusting 10% BHC.

vi. Cotton balls soaked in chloroform or petroleum or kerosine be placed in boring holes and sealed with mud for killing the pest inside the tunnel only.

7a. *Acanthophorus serraticornis* (Fig. 17.6)

It damage crop as above pest.

8. The cerambycid beetle

Dialeges pauper Pascoe

Distribution

Assam, Allahabad, Darjeeling, etc. Outside India, it has been reported from Singapore and Borneo.

Marks of Identification

Beetles are reddish brown to dark brown, measuring from 15 mm to 30 mm in body length and 3.5 mm to 8 mm in breadth. The head of the beetle is elongated posteriorly. Eyes are completely divided. Antennae 1/3 or more longer than body in males and 1/6 longer than body in females. Elytra is truncate at apex, femora is stout in male. Hind tarsus is narrow in both the sexes. Young grub measures about 1/4 in inch and full grown about 1 to 2 inch in body length.

Life Cycle

Beetles appear on the wing in April. Female lay her eggs on the bark of the tree. Eggs hatched in short period. Hatched grubs start feeding on bast by making galleries. Then they bore into sapwood and prepare irregular galleries. Full grown grubs prepare pupal chamber in sap wood in the longitudinal axis of the tree. The full grown grub pupates in pupal chamber. Matured beetles leave the pupal chamber and excavate from the trees for further generation. Two generations are completed in single year.

Nature of Damage

Damage is caused by grubs to the standing trees and felled trees by making galleries and chambers in sapwood region and thus affecting the quality of wood material. Damage to standing plants affect the vigour and growth of the plant.

Host Plants

Sal *Shorea robusta*

Control Measures

i. As suggested for above pest.
ii. Use *Bracon* sp (Hymenoptera : Braconidae) against grubs of this pest since it cause mortalities in them.

9. The sal weevil

Conarthrus jansoni Woll

Distribution

Assam, Goal para, Kachugaon, etc.

Marks of Identification

The weevil measures about 6 mm in body length. It is shining black with long rostrum which is swollen anteriorly. Its prothorax is elongate and convex and brodest at middle. Legs are ferruginous brown coloured. Elytra is as wide as broadest part of prothorax. Under surface of the weevil is black and shining.

Life Cycle

The female lay her eggs in a ovipositing chamber by tunnelling the sapwood, the egg galleries. Grubs feed on making galleries in the stem. They pupate in pupal chamber made in heart wood region of the plant.

Nature of Damage

Damage is caused by both the individuals *viz* grubs and beetles by making galleries into the sapwood and heart wood which are felled or partially dry.

Host Plant

Sal *Shorea robusta*

Control Measures

i. Removal of felled trees from forest.

ii. Treating the felled trees with any of the conventional pesticides which are ecofriendly.

10. The sal scolytid

Sphaerotrypes assamensis Stebbing (Fig. 17.7)

Distribution

Assam and West Bengal

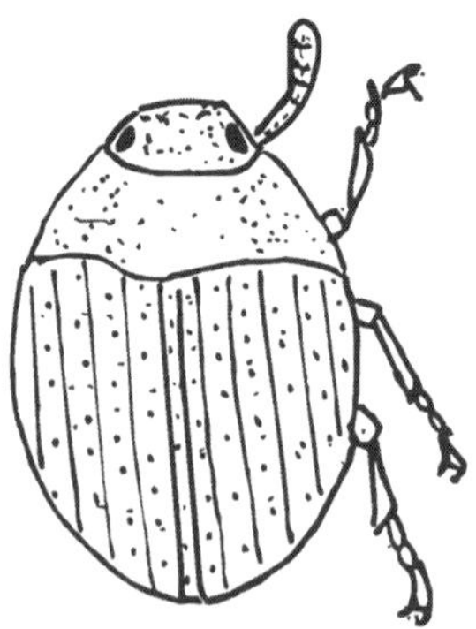

Fig. 17.7 : *Sphaerotrypes assamensis* (Female)

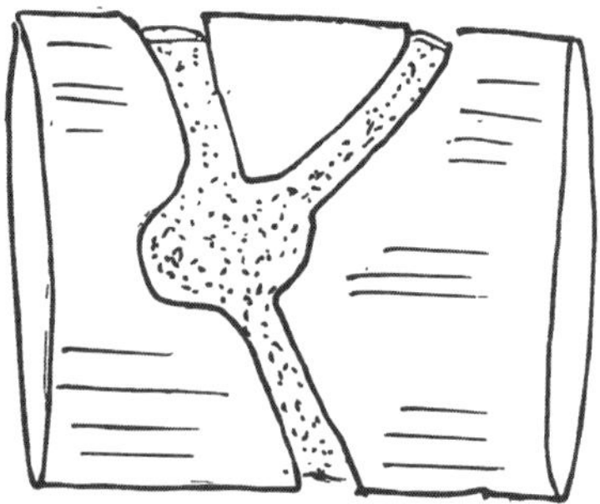

Fig. 17.8 : Mating chamber

Marks of Identification

The beetle measures about 2.8 to 3.6 mm in body length. It is black and slightly shining, prothorax is uniformly punctate and elytra is with deep longitudinal striae, elytral striae are deeper than *S. siwalikensis.* It shows yellow antennae and piceous brown tarsi. Tibiae of this beetle are fringed with white hairs. From under side, the beetle is finely punctate and black. Its larva is white, short and squarish with yellow small head.

Life cycle

Beetles appear in early March and Start egg laying in the bast and sapwood chambers. Mating never takes place outside the mating chamber. Therefore, male after its appearence in forest in March start boring the bark. Upper part of trunk and crown of a large green tree felled in February are badly attacked by the beetles. Males bore into bark of main stem or thicker branches for making mating chamber (Fig. 17.8). It tunnelsout the chamber which can afford space for two beetles as a mating chamber. The female usually enters this mating chamber by boring bark separately from out side. The female is fertilized by the male in this chamber. Later she start boring a separate tunnel through the bark to bast. In bast, she prepares an egg gallery which is about an inch in length. She lay eggs in this gallery. After hatching the eggs, newly emerged grubs start boring tunnels separately as described under *S. siwalikensis.* Grubs prepare winding galleries those increased by the proportion of size of grub. A typical gallery is prepared by the female of this pest. The mated female can lay about 35-40 eggs on each side of the gallery. As much as 70-80 eggs are laid by a single mated female. Both males and females are polygamus. Female mate frequently with the male as per its need, requirement of eggs. Grubs develop in the extension galleries. The full grown grub prepare pupal chamber at the end of the gallery and pupate in it. The pupa moult into adult which find its way outside the tree by making outlet tunnel. There are five to six generations in a year. The second generation is completed by the end of first or second week of June, third at the end of July, fourth in the middle of September and fifth at the end of October. October beetles pair and oviposit as sixth generation.

Nature of Damage

Beetles attack bast and sapwood by making galleries for oviposition. Beetles feed on green cambium. The pest is bast consumer. It also attack young saplings and newly felled trees for oviposition and later for feeding upon them by boring. The pest can also attack sick standing and healthy green standing trees. The infested tree shows a copious outflow of sap that drowns the beetles and kills the eggs and grubs many times.

Host Plants

Sal *Shorea robusta*

Control Measures

i. Trapping the beetles on felled trees and destruction of them.

ii. Use trap trees for collection of beetles after trapping the beetles on trap crop it should be destroyed.

iii. Trees are barked and the bark is then burnt. In the bark and sapwood, larvae, pupae and immature beetles are found. They should be destroyed.

iv. Ecourage to natural enemies

a. *Niponius andrewesi* Lewis predates on grubs of this pest.

b. *Tillioera assamensis* Stebbing feed immature forms of this pest.

c. *Platysoma* sp. : This beetle also feed on grubs and adults of the pest species.

11. The sal xyleborus beetle

Xyleborus major Stebbing

Distribution

Assam

Marks of Identification

The beetle measures about 5.4 to 5.8 mm in body length. It is robust thick and moderately shining. Its head is reddish and large. A fringe of yellow hairs is located at base of mandible. Elytra is dark reddish brown. Larva is white, thick and curved, pupa is whitish yellow.

Life Cycle

Male bore into bark for preparing mating chamber. It travel about half an inch. In the mating chamber male fertilizes two females and some times three. After mating individual females start working directly downward. They lay eggs in egg galleries. After hatching the eggs the grubs bore into sapwood and pupate in it. When full grown and finally when adult is formed it can find its way outside the pupal chamber by outlet tunnel. There are two generations in a year.

Nature of Damage

Beetles bore into green wood of standing trees and sickly trees. The beetle ruins the timber of the tree in severe infestation.

Host Plants

Sal *Shorea robusta*

Control Measures

i. Collection and destruction of beetles.

ii. Treating the crop with 5% Aldrin/Chlordane dust.

12. The sal platypodid

Platypus curtus Chap.

Distribution

Assam, Singapore and Sarawak

Marks of Identification

The beetle is dark brown which measures about 3.5 mm in body length and is robust, short and truncate posteriorly. Apical half of elytra is black and with golden yellow short hairs. Club of antennae is not articulated, prothorax is square.

Life Cycle

Beetles appear in May and start boring the bark of sal tree for oviposition. Beetles bore green trees felled and green sickly standing trees but does not attack the dead wood. The male prepare mating chamber. Female mate with male in this chamber. After mating she starts making egg gallery and oviposit in it.

After hatching the eggs, newly emerged grubs feed on the fungi lining the walls of the egg gallery or tunnel in the wood. Grubs fullgrown and pupate in pupal chamber prepared by full grown larva. The matured adult get escaped from the chamber by out let tunnel prepared by the adult itself.

Nature of Damage

In severe infestation this pest badly pin-holes freshly felled sal trees, freshly sawan logs lying in the forest and standing sickly trees. It attacks mostly green wood.

Host Plants

Sal *Shorea robusta*

Control Measures

i. Collection and destruction of beetles in May.
ii. Barking the felled tree gives protection against this pest.
iii. Felled trees should be removed from the forest as soon as possible.
iv. Encourage the following natural enemy.

Tillioera asamensis predates up on the grubs of this pest.

13. The platypodid beetle

Diapus furtivus sampson

Distribution

Assam, Central provinces

Marks of Identification

The beetle is ferruginous brown or black shining. Its head is punctate with front concavity. Antennae are inserted in a line almost parellel with upper edge of eyes. Elytra is with yellowish patches on the median portion. Thorax is very short and oblong. In female last abdominal segment is deeply concave. The beetle measures about 4 mm in body length.

Life Cycle

Life cycle pattern of this pest is more or less similar to above pest. The pest completes at least two generations in single year.

Nature of Damage

Beetles bore into green wood, standing green sickly trees, and newly felled trees.

Control Measures

i. As per above pest.

ii. *Platysoma* sp. is hesterid which predates on the pest.

iii. *T. assamensis* feed on this pest.

14. The lymantrid caterpillars (Fig. 17.9)

Lymantria semicincta

L. mathura

Lymantria sp.

(Lymantriidae : Lepidoptera)

Distribution

Western Ghats, Assam, Himalaya, UP, Dehradun, Sikkim, etc. Out side India, it is recorded from Japan, USA, several Asiatic countries, Russia, etc. This pest occurs in cool, temperatae to warm climates and in fixed forests, temperate coniferous, tropical and subtropical broad leaf dry forests and moist broad leaf forest.

Marks of Identification

L. mathura : The moth of *L. mathura* (Fig. 17.9) shows black spots on vertex and abdomen being small. The ground colour of fore wing is paler and dorsal side is black zigzag markings are present. Hind wings are orange with black spot at the end of cell and conjoined series of sub marginal spots showing a curved band. The above characters are noticed in males of *L. mathura.* Females are characterized by having head and thorax white, frons fuscus and two black spots on vertex. Abdomen is crimson red with small black spots on vertex. Abdominal terminal segments are whitish. Legs are black and crimson. The wing expanse of male is 40-51 mm and of female 96 mm to 112 mm. Forewings are white with crimson and black spots and with typical zigzag marks. Hind wings are crimson with fuscus sposts at the end of cell. A submarginal maculate band is present on hind wing with some spots on centre of margin. The larva is brownish with full of hairs present on body.

L. semicincta : Females of this species are with black legs, antennae and palpi. Head is yellowish white with black spot behind it. Fore wings black with yellowish white spot at base on inner margin and one in end of cell and three conjoined post medial spots. Hind wings crimson with a broad marginal black border. Wing expanse is about 60 mm.

Life Cycle (Fig. 17.9a, b, c, b)

L. mathura : Mated females lay her eggs on leaves of sal. Hundreds of eggs are laid by a single female in batches or in clusters. Eggs hatch in March/April. Larvae are gregarious feeders. They fee on leaves. The larva shows 5 to 6 instars. The larva period is about 85 days. The pupal period is 10 to 15 days. The full grown larva pupates in soil. Only one generation is completed in a year.

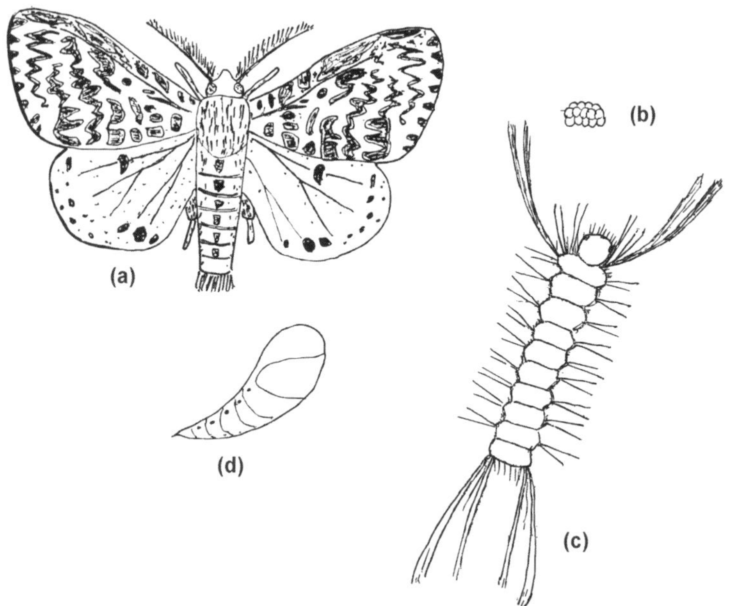

Fig. 17.9 : Life cycle of *Lymantria mathura* : a. male moth; b. eggs; c. larva; d. pupa

Nature of Damage

Caterpillars feed gregariously on flowers and leaves and skeletonize the plant completely in severe infestation. Young shoots are completely destroyed. Hence many times total loss is not uncommon.

Host Plants

This is polyphagus. It damages a very large number of plants belonging to 45 genera of 24 families. Important host plant refers to Sal *Shorea robusta,* Asna *Terminalia tomentosa,* Australian red cedar *Toona ciliata,* Zelkowa *Salix fragilis,* Cherry *Prunnus* sp., Willow *Salix* sp., Peech *Fagus* sp., Apple, Ash, Walnut, Oak, Pine, Cotton wood, Chest nut, Chinaberry tree, Wax tree, etc.

Control Measures

i. Collection and destruction of egg masses.
ii. Collection and destruciton of gregarious and solitary larvae from leaves.
iii. Digging soil for exposing pupae of the pest to natural mortality factors.
iv. Use of pheromone traps for catching and killing moths.
v. Use light traps. The moths attracted to black coloured traps.
vi. Spray the crop with Azadirachtin 0.03 % or DDVP 0.03 %.

15. The pyralid seed caterpillar

Dichrocrosis leptalis

Distribution

Western Ghats, UP, Orissa, MS, Gujarat, TN, AP, etc.

Marks of Identification

The moth is medium sized and orange yellow brown coloured. Forewings and hind wings contain black spots. Larva is reddish brown in body colour.

Life Cycle

Eggs are laid on flower buds/leaves/shoots and developing fruits. Larva bore into terminal shoots and seed capsules. It make webbings and pupate in seed capsule. Many generations are completed in a year.

Nature of Damage

Larvae bore into the shoots and seed capsules and damage developing seeds and capsules.

Host Plants

Sal *Shorea robusta*

Control Measures

i. Collection and destruction of infested plant parts. (seed capsule, shoots, etc.).
ii. Use of biological pest control agents such as *Apanteles* sp., *Bracon* sp., *Diadegma* sp., and *Theronia* sp.
iii. Spraying the crop with DDVP 0.03 % or 0.03 Azadirachtin.

16. The sal leaf caterpillar

Ingura subapicalis

Distribution

Western Ghats, North West Himalayas, West Bengal, Pakistan.

Marks of Identification

Moths are dark fuscous greyish brown. The wing expanse of male is about 28 mm in male 36 mm in female. Antennae of male are with branches on the outerside very short. Forewings with double dark antemedial line incurred below the cell. Remified and very indistinct orbicular spots present and with obsured wavy medial line. Hind wings with basal area pale and with pale patch at anal angle. Antenna is as long as fore wing. Larva is characterized by presence of four paired prolegs.

Life Cycle

Mated females lay their eggs on leaves and flowers of sal tree. They hatch within few days. Newly emerged caterpillars feed on tender leaves and flowers and skeletonize the plant in severe infestation. The larva pass into 5 instars. The full grown larva pupate in soil. Later, the pupa is transformed into adult moth. The moth aftermating lay its eggs on the plant and life cycle is repeated.

Nature of Damage

Caterpillars cause damage to leaves, buds and flowers by feeding upon them.

Host Plants

Sal *Shorea robusta*

Control Measures

i. Collection and destruction of egg masses and caterpillars.

ii. Collection and destruction of moths by using light traps or insect net.

iii. Treating the crop with Azadirachtin 0.03 % or DDVP 0.03% spray.

17. The sal coccid

Monophebus sp.

Distribution

Western Ghats, Himalayas, UP, Assam, etc.

Marks of Identification

Scale insects are semicircular, plate like bodies found remained stick to the plants specialy on leaves and stems. There is sexual diamorphism in this species. Males are winged and females are wingless. Nymphs are similar to female up to certain extent but smaller in size.

Life Cycle

Mated female lay her eggs in ovisac present in her abdominal part. After hatching the eggs the nymphs start emerging from under surface of the female which is already settled on the plant and immovable. The nymphs moult for 3 to 4 times and become the adults. The pest reproduce oviparously.

Nature Damage

Both females and nymphs cause the damage to sal plant by sucking the cell sap. While sucking cell sap the pest inject toxins into plant body which result into curling of leaves, yellowing of leaves, drying of leaves, droping of flowering bodies. The pest also affect the fruit setting adversely. The pest also secrete honey dew like substance on leaves and create sooty moulds over the leaf surface and affect photosynthesis, growth, vigour and yield of the crop.

Host Plants

Sal *Shorea robusta*

Control Measures

i. Scrapping the infested stems and leaves with wooden knife and collection and destruction of them.

ii. Use of conventional pesticides : Spray the crop with Rogor 0.03% or phosphamidon 0.03% or DDVP 0.03%.

18

Pests of Babul (*Acacia arabica*)

Babul *Acacia arabica* is quite important plant of forest ecosystem. There are about 24 species of the genus *Acacia* in India. These xerophytic plants can be successfully grown in the dry and arid regions of forest. Acacia timber has tremomdous importance in making wooden articles like, door, window, cabinets, cheap furnitures and other material of sports. The leaves and pods are used as fodder. More than 100 species of insect pests are associated with *Acacia* spp in India. The most important insect pests of *A. arabica* are listed in the following table.

Sr. No.	Common Name	Scientific Name	Family	Order
1.	The babul buprestid	*Psiloptera coerulea*	Buprestidae	Coleoptera
2.	The babul cerambycid	*Aeolesthes holosericea*	Cerambycidae	Coleoptera
3.	The babul root borer	*Coelosterna scabrata*	Cerambycidae	Coleoptera
4.	The lymantrid caterpillar	*Calliteara* (*Dasychira*) *grotei*	Lymantriidae	Lepidoptera
5.	The Lymantrid moth	*Euproctis scintillans*	Lymantriidae	Lepidoptera
6.	The babul aphid	*Aphis* sp.	Aphididae	Hemiptera

1. The babul buprestid

Psiloptera coerulea Oliv.

Distribution

Karnataka, Pondichery, Maharashtra, MP, UP and Several other states. Sri Lanka.

Marks of Identification

The blue or blue green beetle measures about 15 mm to 25 mm in body length and 5 mm to 10 mm in breadth. It has green head. The beetle is typically buprestid type. Medianly the elytra is with golden red or crimson band. A triangular yellow patch is located at lateral upper corner of abdominal segments. The beetle has green legs with tarsi bronze green.

Nature of Damage

The pest bore the stem in monsoon season.

Control

i. Collection and destruction of beetles.

ii. Dusting crop with 5 % Aldrin/Chlordane.

2. The babul cerambycid

Aeolesthes holosericea Fabr.

Distribution

AP, UP, MS, Assam, North West India, Andaman & Nicobar, Sri Lanka, etc.

Marks of Identification

The beetle measures about 25 to 35 mm long. It is dark brown or reddish brown. Elytra is with light coated silk. Its head is triangular. Prothorax is rounded at sides. Grubs are whitish with brown head and black mandibles.

Life Cycle

Mated female lay her eggs on the wounds or on the joints of two branches of babul in May-June. On hatching the grub feed on inner layers of thick bark, making galleries. They also travel into the sapwood and pupate there. Larval period is about 6 to 9 months and pupal period is 6 to 8 weeks. The beetle can extend several months in pupal stage in chamber. Only one generation is possible in a year.

Host Plants

Terminalia tomentosa, Sal *Shorea robusta,* Mango *Mangifera indica,* Babul *Acacia arabica,* etc.

Nature of Damage

Grubs bore into the stem. Thus, affect the growth of the plant and quality of timber adversely.

Control Measures

i. Collection and destruciton of beetles when they appear in May/June and at the time of oviposition.

ii. Cotton ball soaked with petroleum/chloroform/kerosine be placed in bored tunnel and sealed with mud for killing the pest inside the tunnel.

3. The babul root borer

Coelosterna scabrata Fabr. (Fig. 18.1)

Distribution

Maharashtra, Western Ghats, Oudh, etc.

Marks of Identification

The beetle measures about 1.25 mm to 1.50 mm in body length. The beetle is dull yellowish brown; Its legs are bluish and sides of the body are bluish. The prothorax is with spiny pronotum and with two black spots. Scutellum is large and bluish. The elytra is smooth, broader than prothorax at base. The grub is elongated with brown head and black mandibles. Pupa is pale reddish yellow.

Life Cycle

Beetles appear in October and September. The mated female lay her eggs on the stem in October. On hatching the eggs, the grubs tunnel down to the stem and feed upon woody content. As grub grows older, it travels downward by boring the stem and eating the bored content. Later, it travel down to the base of the stem or root region by feeding on woody content. The larval period is about 6-8 months in November to July. The full grown grub pupate in Pupal chamber. The grubs block the pupal chamber with fibrous material other wise the tunnel is opening outside. The pupal period is 4 weeks i.e. July to August. During the first week of October, the beetle matures and emerges out through the fibrous material of the exit tunnel. In September immature beetles are seen in roots. Only one generation is completed in a year.

Nature of Damage

Grubs cause damage by tunnelling down the centre of stem and later the thicker portion of the stem, the base of the stem and then the roots. Thus, cause the death of the tree. The pest is also responsible for affecting the quality of wood by boring it.

Host Plants

Sal *Shorea robusta,* Casuarina *Casuarina equisetifolia,* Babul *Acacia arabica,* etc.

Fig. 18.1 : *Coelosterna scabrata* (Female)

Control Measures

i. Collection and destruction of beetles in the month of September-October when they freshly emerge from pupal chambers.

ii. Filling wounds or bored portions with cotton ball soaked with petroleum oil or chloroform or kerosine and plugging the same with mud will kill the pest inside the tunnel.

iii. Use conventional insecticides for treating the crop against the beetles.

iv. Collection and destruction of infested plant parts along with pest stages.

4. The lymantrid caterpillar

Calliteara (*Dasychira*) *grotei* Moore

Distribution

Western Ghats, Maharashtra, Karnataka, MP, T.N. & UP.

Marks of Identification

Male moths are stout, active and with light brown fore wings. Their hind wings are yellowish brown and the wing expanse is from 38 to 40 mm while, females have from 60 to 86 mm. The female has bulky abdomen. The full grown larva measures about 50 mm. It is yellowish with rows of brushes of long white hairs. On first four abdominal segments there are four dorsal cones of yellow hairs and black velvety patch between first and second cones.

Life Cycle

The female moth lay her eggs on tender leaves of *Acacia* plant. Eggs hatch within few days. The hatched larvae feed on the leaves. They feed on entire leaf except the hard mid rib and moult for five times. Larvae descend down on ground and pupate in soil or debris associated. The life cycle is completed in about 35 to 40 days. There are about four generations in a year. Thus, the moths appear in January to March, April, June and August.

Nature of Damage

Caterpillars are only destructive. They feed on leaves leaving intact only hard mid ribs on the plant. Thus, affecting plant growth.

Host Plants

It is polyphagus pest. It has many alternative host plants. The host plants of this pest are listed below. *Shorea robusta, Mallotus philippinensis, Terminalia crenulata, Quercus leucotrichophora, Tectona grandis, Psidium guajava, Sizigium aquea, Wagatea spicata, Lagerstroemia flosreginae, Acacia dealbata, A. nilotica,* etc.

Control Measures

i. Collection and destruction of pest stages.
ii. Spray the crop with 0.1 % carbaryl.
iii. Larvae are parasitized by a Microgastrin parasitoid which cause mortalies.

5. The lymantrid moth

Euproctis scintillans Wlk.

Distribution

Western Ghats, MS, Karnataka, TN, and several other states. Andamans, Shri Lanka, Myanmar.

Marks of Identification

The wing expanse of male moth is 20-26 mm and of female moth is 32-38 mm. Fore wings of these moths are uniform vinous brown irrorated with dark scales, the outer margin is yellow but no yellow on disk and without submarginal black spots. The costa is yellow. Hind wings are yellow or fuscus brown with a broad yellow margin. The larva is dark brown with a series of crimson lateral tubercles on a yellow line bearing tuffts of gray hair. Third segment is banded with yellow, short brown hair tufts on 4th, 5th & 11th segments; 5th and 10th segments with broad, dorsal, yellow stripes. Anal segment shows a yellow spot.

Life Cycle

The mated female lay her eggs on the tender leaves in groups or clusters. Clusters are covered with hairs discharged from the body by the female. Eggs hatch within few days. Larvae feed on leaves and moult for 5 times. The full grown larva measures about 30 mm in body length. Larvae are yellow with tufts of dark hairs. The last instar pupate in a thin cocoon prepared on the leaf. The pupal period lasts for 8 to 30 days depending on climatic conditions. Many generations are completed in a single year.

Nature of Damage

Caterpillars cause damage to the trees by feeding and skeletonizing leaves and thus, affecting the growth of the tree.

Host Plants

It is polyphagus pest. Its alternative hosts refers to *Shorea robusta, Ficus bengalensis,* Bamboo, *Cassia fistula, Aesculus indica Acacia decurrens, A. nilotica,* etc.

Control Measures

As above pest

6. The babul aphid

Aphis sp. (Fig. 18.2a) & *A. craccivora.* (Fig. 18.2b)

Distribution

MS, AP, TN, M.P, UP, Bihar, Western Ghats, Karnataka, etc.

Marks of Identification

Aphids are soft bodied insects, louse like in appearence and with two bars or cornicles on the abdomen. They may be winged or wingless. Fore wings are approximately twice the length of hind wings. Nymphs are light yellow and smaller than adults.

Life Cycle

Aphids reproduce parthenogenetically, oviparously and viviparously. Parthenogenetically only females are produced. Hence built up their population within short time.

Nature of Damage

Both nymphs and adults suck the cell sap from tender parts of the babul. They secrete honey dew like substance, create sooty moulds, affect photosynthesis, growth and vitality of the plant.

Host Plants

It is polyphagus pest.

Control Measures

i. Collection and destrution of infested plant parts along with pest stages.

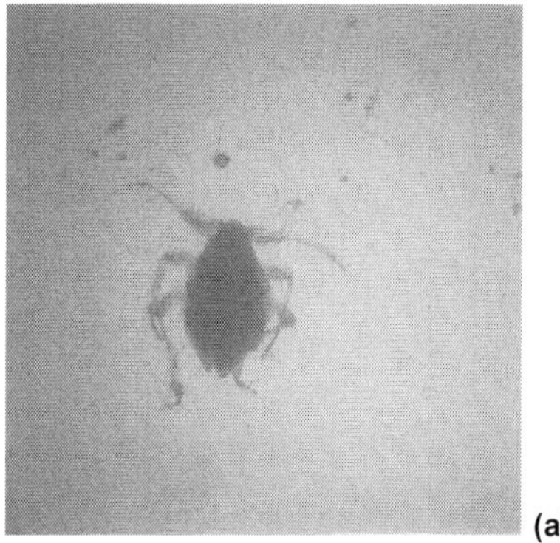
(a)

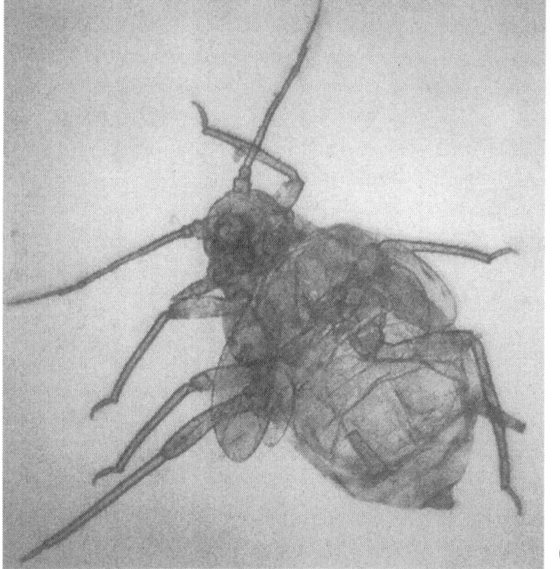
(b)

Fig. 18.2 : (a) *Aphis* sp.; (b) *Aphis craccivora*

ii. Encouragement to predators like ladybird beetles.

iii. Chemical control

Spray - Rogor 0.03% or

Phosphamidon 0.03% or 0.03% DDVP or

Azadirachtin 0.03%

7. Acacia catechu weevil

Myllocerus catechu Marshall (Fig. 18.3)

Distribution

Maharashtra, Pune, Kolhapur

Marks of Identification

The weevil measures about 2.5 mm in body length and 1 mm in breadth. It is black with uniform pale metalic-green scaling. Its eyes are lateral and rostrum is little longer than head. Legs are flavescent with femora green scaling and with minute tooth. Its antennae are ferrugenous.

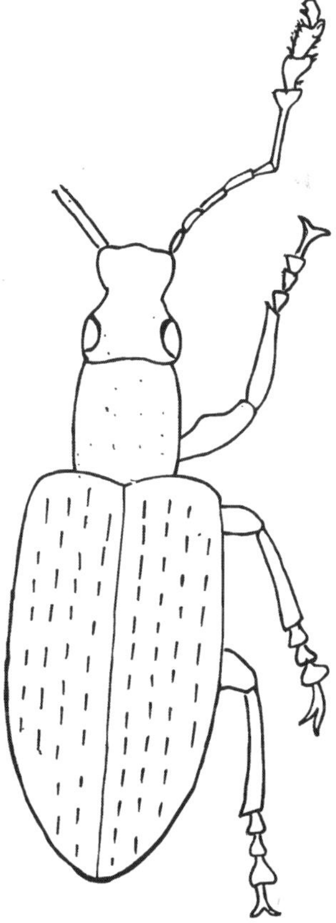

Fig. 18.3 : *Myllocerus catechu*

Life Cycle

Mated females lay her eggs in soil. The grub on hatching from the egg feed on under ground parts and pupate in soil and become adult. The adult is the only destructive form of pest to foliage of *A. catechu.*

Nature of Damage

Weevils feed on leaves of *A. catechu* and affect the growth and vitality of the crop.

Host Plants

Acacia catechu

Control Measures

i. Collection and destruction of weevils.

ii. Ploughing or digging the soil for exposing the life stages of pests to natural mortality factors.

iii. Dusting the crop with 5% Aldrin or 10% BHC.

iv. Spraying the crop with 0.05% DDVP or 0.04% quinolphos.

19

PEST OF SUBABUL (*LEUCAENA LEUCOCEPHALA*)

Subabul *Leucaena leucocephala* is very important plant of Indian forest and plains. It is native of Cuba and reported from Indonesia, Taiwan, Thailand, etc. It has fodder and timber value. It is widely grown in India. This plant is attacked by dozen of insect pests. Important insect pests are listed in the following table.

Sr. No.	Common Name	Scientific Name	Family	Order
1.	The Subabul psyllid	*Heteropsulla cubana*	Psyllidae	Hemiptera
2.	The Subabul scale	*Aspidoitus latinae*	Coccidae	Hemiptera
3.	The scolytid beetle	*Xylebours morigerus*	Scolytidae	Coleoptera
4.	The pod/seed borer	*Araecerus fasciculatus*		Lepidoptera
5.	Termites			

1. The subabul psyllid

Heteropsylla cubana (Figs. 19.1 & 19.2)

Distribution

Throughout India, Native of Cuba; Carribean Island, Phillipines, Indonesia, Taiwan, Thailand, Sri Lanka, etc.

Marks of Identification

The psyllid is cell sap sucking insect which measures about 0.32 mm in body length & 0.15 mm in width in adult stage. Nymphs resmbles with adults except they are smaller in size and without wings. Eggs are whitish to yellowish and elongate. Adults are light yellow with 4 to 5 antenal segments and numerous setae and

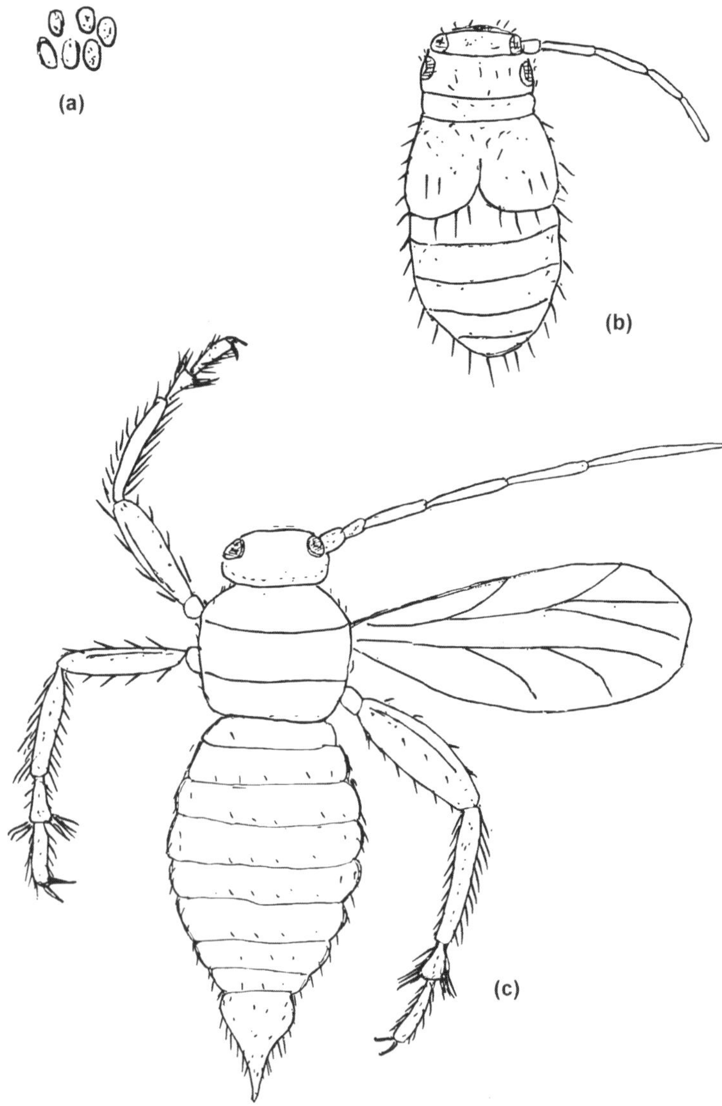

Fig. 19.1 : Subabul Psyllid (*H. cubana*) : Life cycle : a. eggs; b. nymph; c. adult

spines on abdomen & thorax. Legs are short with spines. Fore wings are 0.8 mm long.

Life Cycle (Fig. 19.1a, b, c)

Females lay her eggs on tender leaves of Subabul twigs. They hatch within 2-3 days. Newly emerged nymphs wander here and there in search of tender portions of the plant and start sucking the cell sap from tender leaves. There are five instars in the nymphs. The nymphal period is 8 to 10 days. The life cycle is completed within 10 to 13 days. Adults can survive for 7 days. During a year as much as 8 to 10 generations are completed on subabul.

Nature of Damage (Fig. 19.2a, b)

Both, nymphs and adults are destructive to Subabul. They suck the cell sap from tender portion of the plant specially leaves, buds, stems, flowering and fruiting bodies. The psyllids inject toxins into the plant body when they suck the cell sap which results in curlin of leaves, yellowing of leaves, droping of flowering bodies, etc. The nymphs and adults secrete honey dew like substance on the leaves and other plant parts which cause sooty moulds on the plant. The sooty moulds on leaves affect the photosynthesis of the crop plant and affect the growth and quality of plant.

Host Plants

Subabul *Leucaena leucocephala*

Control measures

i. Collection and destruction of infested plant parts along with pest stages.

ii. Encourage following natural enemies of the pest.

a. A Lady bird beetle, *Menochilus sexmaculatus* (Coccinellidae - Coleoptera) feed on nymphs and adults of psyllids.

b. *Crysopa* sp. (Neuroptera) feed on nymphs and adults of the pest.

c. Spiders feed on the pest stages.

d. Fungi - *Beauveria bassiana* cause the white muscardine disease in pest and cause death. (it controls about 64 to 100 % of pest population).

**Fig. 19.2 : (a) Twig of subabul infected with psyllids (*H. cubana*)
(b) Leaf of subabul infected with psyllids (*H. cubana*)**

e. Use pest resistant varities such as K6, K8, K500, and K636.

iii. Spray the crop with Malathion 0.03% or Rogor 0.03% or DDVP 0.03% or phosphamidon 0.03 or carbaryl 0.15% or Monocrotophos 0.05% or quinolphos 0.05% or Endosulphan 0.05%.

20

Pest of Satinwood (*Chloroxylon swietenia*) & Other Plants

Satin wood (*Chloroxylon swietenia*) is grown in Tamil Nadu, Sri Lanka, Myanmar, Java, Sumatra, Borneo, etc. It has timber value and widely used in making wooden articles of which has export value. This crop is attacked by following insect pests.

1. The satin wood borer

Aeolesthes induta Newman

(Cerambycidae : Coleoptera)

Distribution

Tamil Nadu, Sri Lanka, Myanmar, Java, Sumatra, Borneo, etc.

Marks of Identification

The beetle is dark brown or reddish brown in colour which measures about 26 mm in body length and 10 mm in breadth. The burrow crossing the under-side of the head between the cheeks is very clearly and strongly marked. In males antennae are very much longer than body while in females they are shorter than body.

Life Cycle

Beetles appear in March and lay eggs in the bark of newly felled or injured trees. Good number of eggs are laid on a single plant. On hatching the eggs, newly emerged grubs start boring a tunnel in between the bark and sap wood. The grubs completely ring the tree by tunnelling. The full grown grub bores a hole into sapwood or some times in heart wood for pupation. The grub prepare pupal chamber with the wood and calcarous substance . The pupal form lasts for until the end of February to March. The pest completes only one generation in a year.

Nature of Damage

Grubs bore into the sapwood making tunnels and there by affecting the growth of the plant and quality of wood.

Host Plants

Satin wood *Chloroxyolon swietenia*

Control Measures

i. Collection and destruction of beetles when they emerge in March.
ii. All satin logs felled for sale should be barked.
iii. Quick removal of sickly trees from forests.
iv. Thinning of trees should be done frequently.
v. Cotton balls soaked with chloroform/petroleum/kerosine be placed in bored tunnel and sealed by the mud for killing the pest stages in side the tunnel.

2. The chermid (The spruce gall aphid)

Chermes abietis

Chermidae (Adelgiade)-Hemiptera

Distribution

USA, Canada

Marks of Identification

Chermids are very similar to aphids. However, they differ from aphids by having no cornicles on the abdomen and both parthenogenetic and sexually perfect females by eggs. Their bodies are covered with flucculent mass of white waxy threads. Both winged and non winged forms are present in chermid.

Life Cycle

The life cycle of chermids is more complicated than aphids. For completing the sexual life cycle chermids required alternate hosts, a fir, douslas fir, larch or pine. Chermid pass through the asexual life cycle year after year on either the primary host spruce or the secondary host.

Host Plants

The pest is polyphagus feed on spruce, fir, larch, pine, etc.

Nature of Damage

Both nymph and adult cause the damage to above host plants by sucking the cell sap from tender parts of the plants and producing sooty moulds on leaves and affecting photosynthesis. This also cause damage by forming the galls on leaves and shoot.

Control Measures

i. Collection and destruction of infested plant parts along with pest population.
ii. Encouraging natural enemies.
iii. Spraying the crop with 0.15% Carbaryl, or 0.03% Malathion or Rogor 0.02% or 0.03% DDVP or 0.03% Phosphamidon.

3. Periodic cicada

Magicicada septendicium (Fig. 20.1)

Cicadidae - Hemiptera

Distribution

USA, India

Marks of Identification

The adult cicada is dark brown to black in colour. It has red eyes. Its wings are folded over the back like a tent. Its head and thorax are broad and abdomen is tapering. The female shows strong ovipositor for egg laying in plant body by making slit. Nymphs are small sized, without wings but heavy bodied with broad abdomen and with powerful legs.

Life Cycle

The adult female cicada lay her eggs on the twig of the tree with the help of strong ovipositor, the female makes the slit in the twig for laying her eggs. She lay her eggs in such slits. Egg hatching period in one week. Newly hatched nypmphs drop down on the ground and burrow into the soil at the depth of 2 ft. Each nymph prepare a cell for enjoying consealed life for longer period, 17 years. Nymphs feed on cell sap of small roots which associated with their living cells or living chambers. Nymphs have indefinite number of instars. Nymphs enjoy sedentary life in the chamber. The last instar nymph come out from their chamber, climb up on the tree and become the adult. The adult mates soon and the mated female start laying her eggs on the tender stem by making slits. The cicada lives for 17 years. It has overlapping broods.

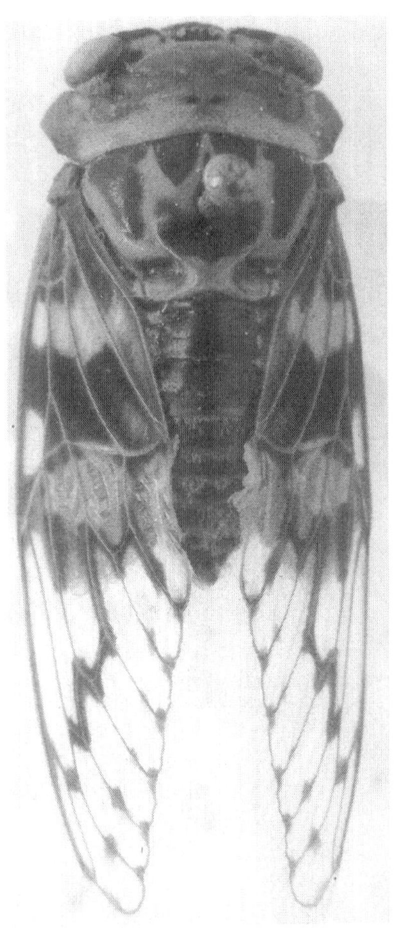

Fig. 20.1 : Cicada

Nature of Damage

Nymph suck the cell sap from roots and inject toxin into the plant body and affect the growth of the tree. While the adults cause the damage by oviposing on tender stems by making numerous slits.

Host Plants

Many hardwood trees in forest.

Control Measures

i. Digging the soil under trees for exposing nymphal chambers and nymphs for natural mortality factors.
ii. Burning all twigs infected with eggs, nymphs and adults.
iii. Use predatory birds like black bird and others.
iv. Use skunks, moles and hogs as predators of nymphal forms.

4. The black pine leaf scale

Aspidoitus sp.

Coccidae - Hemiptera

Distribution

USA, Canada, India, Lake states, California forest.

Marks of Identification

The scales are rounded or semirounded plate like bodies found remain attached to the leaves. Their nymphs are smaller than adult female. Males are winged and females are wingless.

Life Cycle

The scale reproduce viviparously and oviparously. The female gives birth to the nymphs and nymphs moult into adults within 3-4 instars.

Nature of Damage

Both nymphs and adults suck the cell sap from leaves. They settle on the leaves and inject beak into tender leaves and suck the cell sap. While doing so they inject the toxins into the plant body which results in curling the leaves, paleing and drying the leaves. Thus affecting the growth of the tree. They also secrete honey dew like substance which create sooty moulds on the leaves and affect photosynthesis and growth of the tree.

Host Plants

Pines and many other forest trees.

Control

i. Collection and destruction of twigs infested with this pest stages.
ii. Use of natural enemies like lady bird beetles, lace wings etc.
iii. Use of ecofriendly pesticides : Spray DDVP or 0.03% or phosphamidon 0.03% or Carbaryl 0.15%.

5. The cottony maple scale

Pulvinaria innumerabilis

Coccidae - Hemiptera

Distribution

USA, Canada, India.

Marks of Identification

The insect is flat bodied covered with white powdery waxy threads. It is broadly ovate measuring about 1/8 inch in body length. Their eggs are covered with white cottony waxy material and found attached to leaves and othe tender parts of the plant.

Life Cycle

The female lays tremendous number of eggs from 500 to several thousands beneath the white covering of wax (egg sack). Some species lay beneath their arched bodies, some produces young ones. The nymphs come out in July/August and settles on the leaves under sides or along the veins for sucking the cell sap. They feed until the fall. Mature nymphs develop into winged males rapidly than females while, the wingless female emerge later the males. Fertilized females then settle on twigs or small branches of host plant for sucking cell sap. However, she pass the winter and then start laying her eggs on host plant and dies at the settled point.

Nature of Damage

The pest attacks soft mapples. Both, nymphs and adult females suck the cell sap from the plants by injecting toxins into the plant body resulting curling, drying and dropping of leaves from the tree and ultimately affecting the growth of tree and finally the quality of useful product.

Host Plants

Mapples

Control Measures

i. Collection and destruciton of infested plant parts along with pest stages.

ii. Use ecofriendly pesticides : 0.03% Phosphamidon or 0.03% DDVP.

iii. Encourage natural enemies.

iv. Scrapping the bark or infested plant parts with the help of wood knife.

21

PESTS OF NEEM (*AZADIRCHTA INDICA*)

Forest Thrips

Thrips are small insects measuring about 1 to 2 mm in body length and they are always with flowering bodies or tender leaves and suck the cell sap and affect the growth of plants. Important thrips of forest trees are given below.

Sr. No.	Common Name	Scientific Name	Family	Order
1.	Terminalia thrips	*Mallothrips indicus*	Thripidae	Thysanoptera
2.	Babul thrips	*Thilakothrips babuli*	Thripidae	Thysanoptera
3.	Neem thrips	*Heliothrips haemorrhoidalis*	Thripidae	Thysanoptera
4.	Thrips	*Taeniothrips longistylus*	Thripidae	Thysanoptera

1. Neem thrips

Heliothrips haemorrhoidalis (Fig. 21.1)

Distribution

Throughout India

Marks of Identification

Female thrips are slender bodied 1.5 mm long with brown to dark black body colouration. They are pale at the base of abdomen. Thrips are with two pairs of narrow wings which are fringed.

Nymphs are whitish when freshly formed but turn greenish white later and becomes brownish approaching last instar. Nymphs are very small and without wings. They resembles with adults but are miniatures. Nymph shows red eyes. Eggs are been shaped or kidney shaped. Pupal stage is also present in this insect. It is yellow in colour.

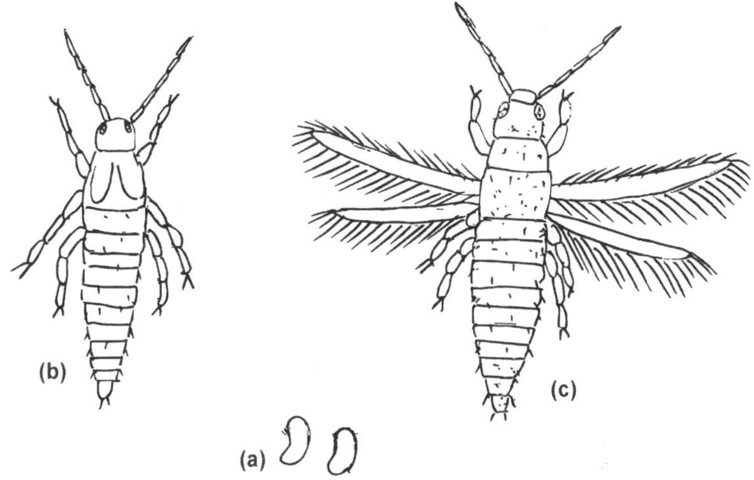

Fig. 21.1 : Thrips : Life cycle : a. eggs; b. nymph; c. adult

Life Cycle

Bean or kidney shaped eggs are laid in leaf tissue. Infact they are injected into leaf tissue singly and covered with a drop of excreta. Incubation period is 2-5 days. Nymphal period is 24 to 26 days in warmer climate and 34 days in cold months pupal stage lasts for 4 days. Thus life cycle is completed within 30-32 days in warm months and 40 days in cold months. Thrips reproduce parthenogenetically also.

Nature of Damage

Both nymphs and adults suck the cell sap from the tender leaves. They lacerate tender leaves or buds and feed on oozing saps which affect the growth of plant.

Host Plants

Neem *Azadirachta indica.*

Control Measures

i. Spray the crop with 0.02% Monocrotohos.

2. Taeniothrips longistylus

It is miner pest of neem which feed on tender leaves by sucking cell sap.

3. The babul thrips

Thilakothrips babuli.

Distribution

It is widely distributed in India.

Marks of Identificatin

Adults are slender bodied with two pairs of narrow wings which are fringed. They shows short antennae and short styli for sucking the cell sap. They measures about 1.2 mm in body length. The nymph is still smaller than the adults and is miniature and lacking the wings.

Life Cycle

Its life cycle is more or less similar to above pest.

Nature of Damage

Both nymphs and adults suck the cell sap from tender leaf lets and flowers and affect the growth and yield of the crop.

Host Plants

Babul *Acacia arabica*

Control Measures

Spray the crop with 0.02 % Monocrotophos.

4. Terminalia thrips

Cause the galls on tender leaves and affect the quality of leaves.

5. The neem pierid caterpillar

Eurema sp. (Pierididae - Lepidoptera)

Distribution

Gujarat, Arid and Semi Arid regions of India.

Marks of Identification

The adult is pierid butterfly. Eggs are whitish and spindle shaped. The full grown larva (caterpillar) mesures about 26-30 mm in body length and green in colour with pale lateral stripes on the body. Its head is black coloured and body segments are annulated. Pupa is olive green but turns brown and then black at the time of hatching of butterfly.

Life Cycle

Eggs are laid in clusters on under surface of leaves/rachis/axils/leaf buds. A female can lay about 28 to 100 eggs. Egg hatching period is 12 to 14 days. Early instars niddle the outer epidermis of leaf lets. Leaves are stripped off by the larvae by feeding upon them. The larval period is 22 to 26 days. The pupal period is 11 to 14 days. The pupa is found attached to host plant. The life cycle is completed in about 45 to 54 days. Many generations are possible in a year.

Nature of Damage

Caterpillars feed on leaves and affect the growth by skeletonizing the plant.

Host Plant

Neem *Azadirachta indica*.

Control Measures

i. Collection and destruction of egg masses, larvae and pupae from the crop.

ii. Encourage following parasitoids which cause mortalities in larval stage of the pest.

a. *Euplictrus* sp. (Ichneumonidae - Hymenoptera)

b. *Brachymeria* sp. (Chalcidae - Hymenoptera)

Former is larval and later is pupal parasitoid.

iii. Chemical control : Spray the crop with 0.02% Monocrotophos.

6. The neem white grub

Holotrichia consanguinea Blanch. (Melolonthidae - Coleoptera)

Distribution

Widely distributed in Western Ghats and Maharashtra and other states of India.

Marks of Identification

The beetle is large sized brownish black in colour and is having typical scarabaeid shape. Its grubs is whitish with strong mandibles of brown to black colour. Head is blackish - brown. It shows 3 pairs of legs. Pupa is exarate type.

Life Cycle

Beetle oviposits in soil at the onset of monsoon in July. After hatching eggs newly emerged grubs feed on roots and humus in soil and full grown in the soil. Pupation takes place deep in the soil. The larval stage lasts for six months and pupal stage 14 days. Adults formed in November pass the winter and summer in soil and emerge in June. Only one generation is possible in a year.

Nature of Damage

Grubs damage roots of neem. While adults feed on foliage at night only. At day time pest remain in the soil.

Host Plants

Neem and several other forest trees.

Control Measures

Dusting the crops with 10% BHC or spraying the crop with 0.05% Endosulphan or treating soil with 0.5% Aldrin against grubs.

Microbiol Control

Use BP *Bacillus Popillae* against this pest. It cause mortality in 3rd instars of the pest.

Bibliography

Ahmed S.I. 1991. Scope of biological control agents in forest pest management in Arid zone. *Proc. Sem. Afforest Arid Lands,* IAZFR, Jodhpur.

Beeson C.F.C 1941. The ecology and control of the forest insects of India and the neigbouring countries. Basant Press Dehradun.

Bhasin, G.D. and M.L. Roonwal. 1954. A list of insects of forest plants in India and the adjacent countries. *Indian For. Bull.* 17(1), (NS), Entomology, Manager of Publicatins, Delhi. Browne F.G. 1986. Pests and diseases of forest plantation trees.

Chatterjee, P.N. and Mishra M.P. 1974. Natural enemy and plant host complex of forest insects pests of Indian region. *Indian For. Bull.,* 265, 233 pp.

Graham S.A. and Night F.B. 1965. Priciples of Forest Entomology. pp. 417.

Jadhav. B.V. & T.V. Sathe. 2006. Biodiversity of Aphids (Order: Hemiptera) from Poona District of Western Maharashtra. *J. Adv. Zool. 27,* 43-45.

Heppner J.B. 1991. Tropical lepidoptera: Faunal regions and the diversity of lepidoptera. *Trop. Lepidoptera* 2 (suppl.-1), 11-85.

Khan H.R., Parasad L. and Sushil Kumar 1985. Some Important Insect Pests of Madhya Pradesh and Their Control. Paper presented in Forestry Conference (M.P.) SFRI, Jabalpur 18-20 Feb. 1985.

Mathur, R.N. 1960. Pests of teak and their control. *Indian forest Res. (NS).* Entomology 10(b)1960 (1961), 43-66.

Monnier, M.F. 1958. Termites of *Eucalyptus. Proc and Conf. for Interfric Pointe* Noire 2, 281-284.

Nair, K.S.S. and G. Mathew 1993. Diversity of insects in Indian forests-The state of our knowlege. Hexaposda, 5(2), 71-78.

Roonwal, M.L. G.D. Bhasin and G.D. Pant 1950. A systematic catalogue of the main identified entomolgical collection of, at the Forest Research Institute, Dehradun. Parts-1 to 3 *Indian Forester,* 76 (1), 498-505.

Sathe, T.V. 1992. Fauna of aphids on plants of economic importance found in Western Maharashtra, India, *J. Curr. Biosci.,* 9(1), 27-30.

Sathe T.V. 1998. Sericultural Crop Protection. APO, pp.1 to 121.

Sathe T.V. 2004. Vermiculture and organic farming. Daya Publishing House. Delhi, pp. 1-22.

Sathe T.V. 2004. Biology of Bracond Pest Biocontrol Agents from Western Maharshtra. *Bull. Biol. Sci, 2,* 73-75.

Sathe T.V. 2004. Biology and Behavior of Coccinellid, beetles, In : Indian insect predators in biological control. Sahayaraj, DPH, Delhi. 177-198.

Sathe T.V. and Bhoje. 2000. Biological Pest Control. DPH, New Delhi. pp. 1-122.

Sathe T.V. and Pandharhate A.R. 1990. Hawk moth (Shpingidae: Lepidoptera) Diversity in Western Maharashtra including Ghats. *Geobios,* 18, 77-82.

Sathe T.V. & A.R. Pandharbale. 2005. Biodiversity of moths (order-Lepidoptera) from Western Ghats of Satara District, *India. Bull. Biol. Sci.,* 1, 81-88.

Sathe T.V. & Y.A. Bhosale. 2001. Insect Pest Predators. DPH, New Delhi, pp. 1-169.

Sathe T.V. & Patil Vaishali J. 2003. Insect Predators and Pest Management. DPH. New Delhi, pp. 1-216.

Sathe T.V. & K.P. Shinde. 2006. Diversity of Butterflies from Western Ghats (Kolhapur district) *J. Nat. Con.,* 81 (1), 181-184.

Sathe T.V. & K.P., Shinde 2005. Biodiversity of Dragonflies from Western Ghats of Maharashtra India. Proc. 4th *Int. Nat. Sym.* Odonatology, Pontendra, Spain. July, 2005.

Sathe T.V., S.A., Inamdar and R.K. Dawale. 2003. Indian Pest Parasitoids. DPH, New Delhi. pp. 1-167.

Sen-Sarma, P.K. 1983. Forest Entomology in India. *Indian Rev. Life Sci.* 3, 89-103.

Sen-Sarma, P.K and M.L. Thakur, 1985. Pest Management in Indian Forestry. *Indian Forester,* 111(11), 956-964.

Singh, P. and R.S. Bhandari. 1988. Insect Pests of Bamboo and Their Control. *Indian Forester,* 114 (10). 670-683.

Stebbing E.P. 1914. Indian Forest Insects of Economic Importance: Coleoptera. Eyre & Spottiswoode Ltd. London, 648 pp.

Thakur, M.L. 1977. Pest Status and Control of Termites in Rural Areas. *Indian Forester,* 106 (6), 425-434.

Troup, R.S. 1921. Silviculture of Indian Trees, Vol. 1, pp. 178-183, Oxford Press.

WRI, IWCN & UNEP, 1992. Global Biodiversity Strategy. WRI, IUCN, UNEP, 244 pp.

INDEX

D

L

M

N

O

P

Q

R